T5-AFS-417

ALLE ZEIT WACH
1842

Emil Supp

How to Produce Methanol from Coal

With 83 Figures

Springer-Verlag Berlin Heidelberg NewYork
London Paris Tokyo Hong Kong Barcelona

Dipl.-Ing. Emil Supp
Lurgi GmbH
Gerviniusstraße 17/19
D-6000 Frankfurt/Main 11
Germany

ISBN 3-540-51923-8 Springer-Verlag Berlin Heidelberg NewYork
ISBN 0-387-51923-8 Springer-Verlag NewYork Berlin Heidelberg

Library of Congress Cataloging-in-Publication Data
Supp, Emil
How to produce methanol from coal / Emil Supp.
ISBN 0-387-51923-8 (acid-free paper)
1. Methanol. 2. Synthesis gas. 3. Coal gasification.
I. Title.
TP358.S83 1990
662'.669--dc20 90-38554

Offsetprinting: Mercedes-Druck, Berlin; Bookbinding: Lüderitz & Bauer, Berlin
2161/3020-543210 – Printed on acid-free paper

Preface

Owing to efforts and legislative action – initiated above all by the government of the United States – to use cleaner fuels and thus make a contribution towards a better environment, public attention is back again on using methanol in carburettor and diesel engines. Most prominent among the raw materials from which methanol can be produced is coal, whose deposits and resources are many times larger than those of liquid and gaseous hydrocarbons.

This book deals with the production of methanol from coal. It describes both the individual steps that are required for this process and the essential ancillary units and offsites associated with the process itself. It is not meant to inform the reader about the intricate details of the processes, which can much better be taken from the specialized literature that deals exclusively and in detail with them or from the well-known standard engineering books. Rather, this book is to give the reader an impression how manifold a field this is, how many process variations and combinations the designer of such plants has to consider in order to arrive at an optimum design in each particular case. Apart from the production of chemical-grade methanol, the book deals briefly also with fuel methanol production, i.e. with the production of alcohol mixes.

One of the many possible routes from coal to methanol is illustrated by a process flow diagram, and a material and energy balance is compiled for this typical example.

No one can trace precisely all the influences which affect a work of this kind. My sincere thanks go to numerous colleagues who have unselfishly assisted me with a number of process steps and with the preparation of this book. I am also indebted to all whose publications I have used to gain information from, and whom I have hopefully all quoted without fail. This book would probably not have been conceived without the support of Dr. Jens Peter Schaefer, Chairman of the Board of Lurgi GmbH, Frankfurt, and it would not have come to an end without a few benevolent pushes from my wife.

October 1989 Emil Supp

Contents

1. How to Produce Gas From Coal

1.1 What and Where Is Coal?

Coal owes its origin almost exclusively to plants that grew on our earth many millions of years ago from the Carboniferous to the Tertiary Period. Vast expanses of forest land slowly subsided so that the forests were covered with water and largely shut off from the oxygen of the air. Various microbiological processes occurring at normal pressures and temperatures turned the dead organic material at first into peat, and later on, as it went further down, into brown coal. After this first coalification stage termed the *biochemical phase* had been completed, the morphologically and chemically inhomogeneous brown coals were converted to bituminous coals in a second, geochemical stage. This second stage, which took place under the long-term effect of an elevated temperature reaching about 150 °C at a depth around 5 000 m, is also called *metamorphosis*. The end product of this development is believed to be graphite.

The great variety of decomposition processes taking place during the coalification phase produced very different types of coal with different chemical, physical and technological properties even if their degree of coalification is the same. Each type of coal contains not only major quantities of intergrown minerals which are mostly removed during coal preparation – e.g. clay, dolomite, pyrite and quartz, to name only a few of them – but also numerous trace elements which range, in alphabetical order, from gold (Au) to zirconium (Zr) and normally account for 0.001 to about 3 wt. % of the coal ashes [1.1]. Not least of all these trace elements – as it will be demonstrated later – are responsible for the difficulties in producing pure synthesis gases.

1.1.1 Reserves and Resources

Bituminous coals and brown coals are abundant almost everywhere on earth. Total reserves are estimated at some 13 000 Gt of bituminous coals and 2 600 Gt of brown coals, of which more than 90 % are anticipated to lie in the northern hemisphere. In spite of the sizable new oil and gas deposits found in recent years, the total coal reserves are still four times as large as the estimated reserves of liquid and gaseous energy sources. They alone would be sufficient to meet the energy demand of a steadily growing world population until long after the turn of the millenium, even though this demand grows beyond proportion. However, only about 400 Gt of brown coals and some 1 200 Gt of bituminous coals are proven.

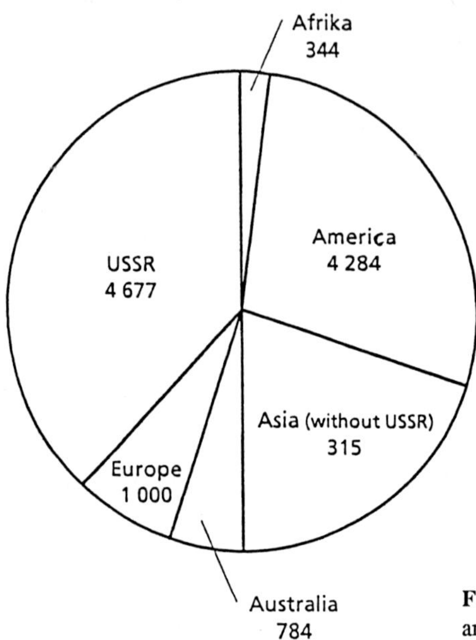

Fig. 1.1. Distribution of the world's coal resources and reserves (figures in Gt)

The current state of mining technology probably does not allow more than 7 to 10% of the proven and anticipated 15 600 Gt to be produced under economically justifiable conditions. The geographic distribution of the anticipated and proven coal reserves is shown in Fig. 1.1. Coal deposits throughout the world, their thicknesses and mining depths, as well as their current and future mining potentials are dealt with in many detailed reports [1.2,3].

1.1.2 Properties of Different Types of Coal

Coals are products of nature with very different compositions and properties. Most countries with rich coal deposits have developed their own classification systems for bituminous coals, the better known ones among them being the 1956 ECE classification according to DIN 23003 and the classification described in ASTM D 388-66 used in the USA. The latter classifies bituminous coals into four coalification ranges termed Bituminous I, Bituminous II, Sub-bituminous II and Lignite IV. Each of these ranges is then further broken down into groups according to volatile contents or calorific value. A uniform classification exists for brown coals. It is described in ISO 2950-1974-(E) and distinguishes between six classes defined according to the water content of the raw coals, which are in turn subdivided into five groups on the basis of tar contents. Coals can be quickly characterized with respect to their technological properties by looking at their moisture, ash and volatile contents. The latter also indicates the degree of coalification; it defines the weight of the gas, tar and water that is lost if the coal is heated to 900 °C.

The main chemical components of all coals are carbon, hydrogen and oxygen, but they also contain nitrogen and sulfur. As the degree of coalification increases, the carbon content will increase and the hydrogen and oxygen contents decrease. Brown coals are characterized by particularly high contents of oxygen – as much as 30 % of the moisture- and ash-free carbon substance (maf) in the case of soft brown coals –, whereas anthracite contains 92 wt. % or more of carbon. The C/H ratio also increases from 1 to more than 2 as the degree of coalification increases.

In addition to the volatile, ash and moisture contents, and to the ultimate analysis, an assessment of the gasification behaviour of the coal and its effect on the product gas properties also requires a knowledge of several other factors such as

- the swelling index, which describes the extent to which the coal softens as it is heated and defines its caking properties
- the composition of the ash and its softening and melting behaviour
- the calorific value.

Compositions of some of the world's commonest coals have been compiled in Table 1.1.

1.1.3 Exploration, Mining and Coal Preparation

A considerable effort is required to locate and mine the coals and to make them "digestible", so to speak, for the different gasification processes. Without going into greater detail about the numerous operations involved – they are described in the relevant literature [1.1] – the list below will give the reader an idea about the number of individual steps to be taken before the coal eventually reaches the gasifier. Essentially, these steps include

- establishing whether a coal seam is worth mining in view of its structure and thickness, and deciding on the most suitable mining method (shaft or strip mining)
- preparing the deposits for mining; this may take decades and include sinking shafts, developing working levels, and then subdividing the developed deposit into workable sections by opening suitable galleries
- mining the coal from open pits (lignite, young bituminous coal) or slopes (steeply inclined bituminous coal seams), or by room-and-pillar methods (flat seams), including the required lifting and hauling operations
- ventilating the working areas to ensure a largely dustfree supply of well-tempered breathing air for the miners and to extract harmful gases, above all the mine gas (methane) that, when left to accumulate, may lead to the much-dreaded firedamp explosions
- preparing the lifted coal for subsequent upgrading. As a first step, this normally involves prescreening the coal to obtain a certain desirable minimum lump size (usually about 150 mm) for the crusher, removing the unwanted material, crushing the coarse lumps, and producing a steady flow of mate-

Table 1.1. Properties of selected coals

	Proximate Analysis				Elementary Analysis [% wt.] (m.a.f.)					Ash Softening Point [°C]
	Ash	Moist.	Vol.Mat.	C_{fix}	C	H	O	S	N	
Anthracite										
– Pennsylvania	6.9	2.3	3.1	87.7	95.4	2.2	1.0	0.5	0.9	n.a.
– Ruhr (F.R.G)					92.3	3.3	3.4	0.9	0.1	1240
Bituminous										
– Low vol. West Virginia	3.9	3.5	18.2	74.4	90.8	4.8	2.8	0.6	1.0	1500
– High vol. B Kentucky	4.2	7.2	39.8	48.8	80.4	5.7	9.2	2.9	1.8	1600
– High vol. C Indiana	8.7	12.4	36.6	42.3	79.3	5.5	10.7	2.9	1.6	1270
Subbituminous										
– B. Wyoming	3.8	23.2	33.3	39.7	74.1	5.3	18.8	0.5	1.3	1590
Lignite										
– North Dakota	6.6	26.1	27.8	39.5	72.6	4.5	21.8	0.3	0.9	1240
– West Germany	5.0	n.a.	n.a.	n.a.	68.1	5.2	25.2	1.0	0.5	950

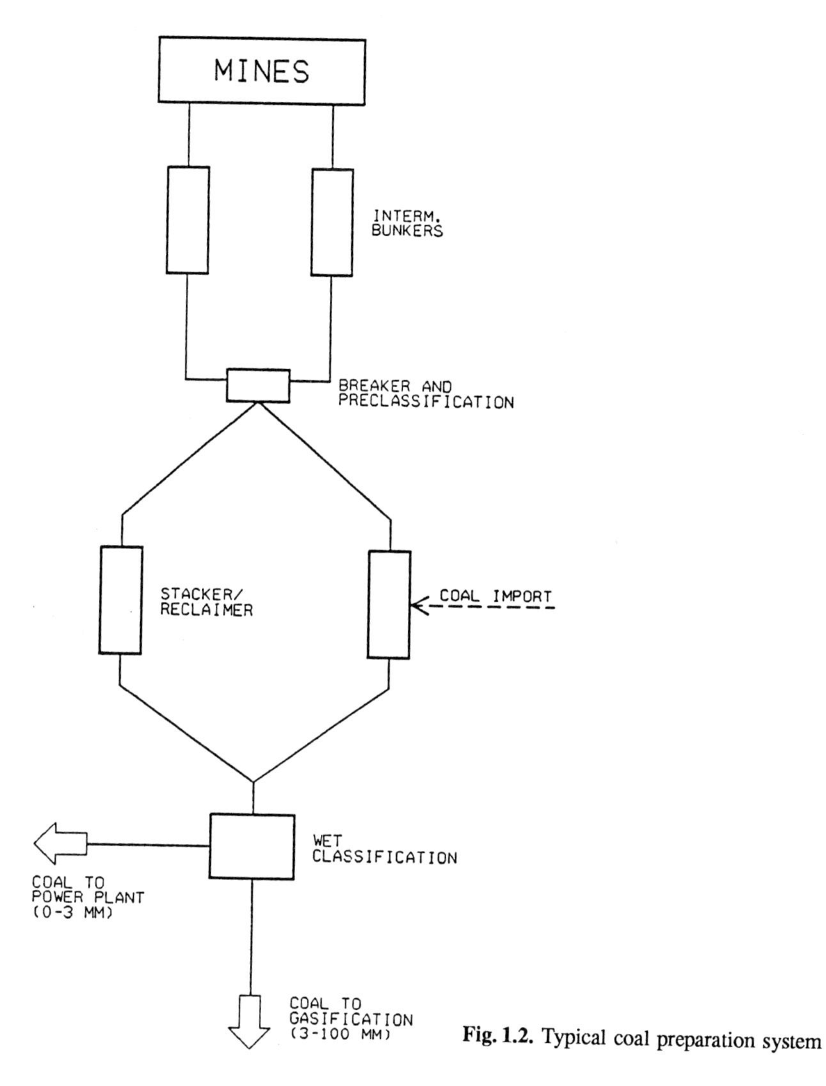

Fig. 1.2. Typical coal preparation system

rial. The latter is achieved by splitting the uneven mass flow up into small quantities which are temporarily stored in different bin compartments and then recomposed into a new and steadier mass flow

- preclassifying the coal to about 10 mm minimum size in a dry system, wet classifying the particles from 150 to 10 mm, and subsequently classifying the under size
- briquetting of fine coals if gasification processes are used which require lumpy coal. To this end, the fine coal is ground down, mixed with a binder (usually pitch) and briquetted in roller presses. This process is usually followed by some thermal aftertreatment, or hot briquetting of caking coals takes the place of cold briquetting with binder
- storing the coal and taking precautions to prevent auto-oxidation, especially in the case of younger coals. This requires either the oxygen of the air to

be kept away from the inside of the stored material (by covering it with fine coal or compacting it), or the heat released during oxidation has to be dispelled by a sufficient exchange of air e.g. by drilling holes into the pile.

It ought to be clear from this brief description that such elaborate coal preparation costs a great deal of money. Large coal gasification plants are therefore erected *on the coal*, i.e. in the immediate neighbourhood of the pitheads, and – wherever possible – are fed with run-of-mine coal so that preparation is limited to grading the coal to a size that is best suited for the selected gasification process. Figure 1.2 presents a schematic illustration of coal preparation facilities in one of the largest syngas production plants using lumpy coal of 3 to 100 mm particle size in fixed-bed gasifiers operating according to the LURGI pressure gasification process.

1.2 Coal Gasification

1.2.1 History

Gas was produced from coal already towards the end of the 18th century, although at the beginning exclusively for firing and lighting purposes. Following low-temperature carbonisation tests in an airtight enclosure (*Minckelen* 1783) [1.1], *Lavoisier* produced the first water gas in 1783. First tests aiming at autothermal gasification of coal, although still with air, began around 1840 (*Faber du Faur*).

Major progress in the production of low-nitrogen gases – which are desirable for almost all syntheses – was not achieved until the twenties and thirties of the present century. This marked the beginning of autothermal gasification with oxygen, as oxygen could now be produced at a competitive price by the *Linde-Fraenkl* process that was developed to industrial maturity at the same time. Processes that gained particular importance were the *Winkler* Fluidized-Bed Process, the LURGI Pressure Gasification Process, and the *Koppers-Totzek* Dust Gasification Process. It was also the time of new allothermal gasification processes producing low-nitrogen synthesis gases, e.g. the *Pintsch-Hillebrand* and the *Koppers* Circulation Process.

After a standstill of almost 20 years, during which coal was largely replaced by oil and natural gas, a new phase in the development began in the USA in the sixties as natural gas was expected to be in short supply in the future. This development was boosted by the first oil price shock at the beginning of the seventies and almost turned into a boom in the USA and in Europe. While in the USA the emphasis was on the production of lean gases for power stations and of substitute natural gas (SNG) for the utility grids, attention in Europe was focussed more on the production of high-H_2 and high-CO synthesis gases. The processes developed in the USA, such as Hygas, Bi-Gas, Cogas, Synthane and others, are therefore designed to produce high-methane gases. A syngas process that is suitable for methanol production was developed by TEXACO and termed high-temperature gasification. This route has recently been enlarged by a two-stage fuelling version suggested by DOW.

In Europe, LURGI and *British Gas Corporation* together developed another variant of the LURGI Pressure Gasification Process, the Slagging Gasifier. Developments also began on entrained-flow processes which are today known by the names SHELL Coal Gasification Process (SCGP) and Prenflo, and the *Winkler* process was further improved to a high-temperature route (HTW).

1.2.2 Methods of Coal Gasification

According to their mechanical technology, i.e. to the method of feeding the coal to the gasifier and to its behaviour in it, coal gasification processes may be classified into

- moving-bed gasification
- fluidized-bed gasification
- entrained-flow gasification
- slurry gasification, a variety of the entrained- flow route.

Going by the way in which the required heat is provided, they may also be grouped into

- autothermal gasification processes, providing the heat required for the reaction by burning some of the coal with oxygen,
- allothermal gasification processes, generating the required heat outside the gasifier and using suitable heat carriers to supply it directly or indirectly to it,
- hydrogenating gasification, which is really a variety of allothermal gasification. As this process route produces high-methane gases (for SNG production), it is not cost-effective for the production of methanol syngas and will therefore be disregarded here.

In addition to the above two criteria, the processes can also be classified according to the types of product gases, e.g. lean gas, reduction gas, rich gas (methane) and synthesis gas. Among these types, however, only the latter needs to be considered here.

1.2.3 Selection of Suitable Processes

Table 1.2 contains a survey of major gasification processes that have proved reliable and safe and are today used for syngas production on an industrial scale. In addition to the operating parameters, the survey also provides data about the suitability of these processes for different types of coals. However, these data have to be taken with a grain of salt. No absolute rating of the coals is possible in view of their wide variety of properties. Future methanol producers will have to make their choice between these processes on the basis of two main criteria – the properties of the available coals and the resulting economics of gasification.

Table 1.2. Coal gasification processes in commercial operation

	Press. Range [bar]	Temp. * [°C]	Anthracite	High Volatile Bituminous	Low Volatile Bituminous	Lignite	Ash >30 %	Fines >30 %
LURGI Dry Bottom	20-100	300-600	1 - 2	1 - 2	1 - 2	2	2	0
British Gas/Lurgi Slagger	20- 70	450	1 - 2	2	2	0 - 1	0	0 - 1
Winkler/HTW	1- 10	1050	1	1	1	2	2	1 - 2**
Koppers-Totzek	atmosph.	1800	1	2	2	2		2
SHELL	20- 40	1500	1	2	2	2	0	2
TEXACO	20- 40	1350	1	2	2	0	0	2
DOW	10- 20	1450/1000	1	2	2	0	0	2

* at outlet gasifier ** max. grain size 6 mm
0 = less suitable; 1 = suitable; 2 = perfectly suitable

Which gasification process will be technically suitable or reasonable has to be decided in each particular case by taking a look at coal properties such as

- volatile contents
- ash contents
- water contents
- reactivity
- caking behaviour
- grindability.

1.2.4 Physical and Chemical Fundamentals of Coal Gasification

Coal gasification occurs according to two different types of reactions: heterogeneous reactions occurring at the phase interface, in which the gasifying agent and the product gases react with the coal, and homogeneous reactions, in which the primary oxidation gases, the gasifying agent, and the devolatilization gases released by heating the coal (pyrolysis) react with each other. It is the interplay of these reactions that determines the composition of the product gas. Depending on the type of gasifier used, these reactions may occur at clearly different times and locations (fixed-bed gasifier) or almost simultaneously (high-temperature dust gasification).

The gasification process may be described by three laws of nature:

- the material balance, in which the mass conservation law requires the masses of the reagents and the products to be equal
- the enthalpy balance, in which the energy conservation law requires the heat input or output during the reaction to be equal to the enthalpy difference between the reagents and the reaction products, irrespective of their changes in the system
- the chemical equilibrium, or the closest possible approximation to it for the chosen method of gasification.

These three factors make it possible to compute the course of gasification theoretically. To this end, numerous equations have to be introduced in an iterative process. However, the way in which gasification is to proceed in practice, and thus the exact compositions, cannot be derived until the kinetic processes, too, are known with sufficient accuracy.

1.2.4.1 Reaction Enthalpies

The complex reaction of coal gasification may be described by the heterogeneous and homogeneous reactions listed in Table 1.3 and by its pyrolysis [1.4]. It will generally be sufficient to have a look at these reactions, since the reaction enthalpies of all other equations can be derived from them by simple arithmetic manipulation. The course of the individual reaction has no influence on the total heat input/output of a reaction sequence as long as the same feedstocks lead to the same products (*Hess's* law).

Table 1.3. Reactions essential for coal gasification

		ΔH *	
1.	Heterogeneous Reactions Gas / Solids		
	– Heterogeneous Watergas Reaction		
	$C + H_2O \longrightarrow CO + H_2$	+ 119	KJ/mol.
	– Boudouard Reaction		
	$C + CO_2 \longrightarrow 2\ CO$	+ 162	KJ/mol.
	– Hydrogenating Gasification		
	$C + 2\ H_2 \longrightarrow CH_4$	– 87	KJ/mol.
	– Partial Oxidation		
	$C + 1/2\ O_2 \longrightarrow CO$	– 123	KJ/mol.
	– Complete Oxidation		
	$C + O_2 \longrightarrow CO_2$	– 406	KJ/mol.
2.	Homogeneous Reactions Gas/Gas		
	– Homogeneous Watergas Reaction		
	$CO + H_2O \longrightarrow H_2 + CO_2$	– 42	KJ/mol.
	– Methanation		
	$CO + 3\ H_2 \longrightarrow CH_4 + H_2O$	– 206	KJ/mol.
3.	Pyrolysis Reactions		
	– $C_1H_xO_y = (1-y) . (C + y) . (CO + x/2) . H_2$	+ 17.4	KJ/mol.
	– $C_1H_xO_y = (1-y - x/8) . (C + y) . CO + x/4 . (H_2 + x/8) . CH_4$	+ 8.1	KJ/mol.

* ΔH at 298 K, 1 bar

To compute the distribution of the main coal components in the product gas, i.e. carbon, hydrogen and oxygen, the simultaneous side reactions are normally neglected in the first run. If the distribution of secondary components such as sulfur in H_2S, COS, mercaptanes etc. or of nitrogen in NH_3 or HCN needs to be known, it can be derived by secondary computations using the distribution of main components computed before.

Table 1.3 illustrates that, among the five primary reactions, the two reactions of hydrogen with carbon dioxide are endothermic while the reactions of hydrogen with oxygen are exothermic. It follows from equations 1 and 5 that 0.29 kg of carbon would have to be burnt to CO_2 to produce the heat required to gasify 1 kg of C to CO and H_2. Important for methanol synthesis are above all the heterogeneous and the homogeneous water gas reaction – which are both significantly influenced by gasification steam – as well as the oxidation processes determining the oxygen demand of the gasification system. Methane formation according to equations 3 and 7 is undesirable.

1.2.4.2 Thermodynamic Equilibrium

Thermodynamic equilibrium makes it possible to compute how far a chemical reaction can proceed under the given conditions of pressure and temperature. For such reversible reactions, the velocity of forward and backward reactions is the same, although thermodynamic considerations provide no clue as to how fast this equilibrium is reached.

Tendencies indicating how pressure and temperature influence the equilibria can be recognized from *Le Chatelier's* and *Braun's* principle that a system reacts to changes in conditions in such a way as to minimize the effects of the applied force. Hence, an increase in pressure shifts the equilibrium towards a smaller volume, whereas a reduced pressure leads to a larger volume. A typical case in point is the methanation reaction shown in Table 1.3, which for the purpose of computing the thermodynamic equilibrium may be rewritten as follows:

$$k_p = \frac{p_{CO} \cdot p^3_{H_2}}{p_{CH_4} \cdot p_{H_2O}} = \frac{y_{CO} \cdot y^3_{H_2}}{y_{CH_4} \cdot y_{H_2O}} \cdot p^2 = f(t).$$

Any increase in pressure will lead to more methane and vice versa.

A temperature increase in an endothermic reaction will increase the quantity of reaction products. A typical example is the heterogeneous water gas reaction in which a higher temperature favours the formation of CO and H_2. In an exothermic reaction, a temperature increase has the opposite effect as shown by the equilibrium behaviour of the homogeneous water gas reaction.

The dependence of the equilibrium constants on temperature may be described by the following equation:

$$\log k_{p,T} = -\frac{H_{298}}{19.146T} + \frac{S_{298}}{19.146} + \frac{a}{19.146} \cdot \left(\ln\frac{T}{298} + \frac{298}{T} - 1 \right)$$

where $a = \sum \nu_i c_{pi} \cdot (T)$ and S_{298} is the reaction entropy, which – similar to the reaction enthalpy – can be computed from the standard figures of the individual components [1.4]. The equilibrium constants of the heterogeneous and homogeneous reactions shown in Table 1.3 as a function of temperature are illustrated in Fig. 1.3. When plotted as $\log k_p$ against $1/T$ for a temperature range as it is relevant for practical applications, the result is a straight line. This way of plotting the function makes the differences between exothermic and endothermic reactions with respect to the behaviour of their equilibria particularly clear. Their tendency depends on the sign and size of the reaction enthalpies. Exothermic reactions are characterized by a positive and endothermic reactions by a negative tendency. The greater the reaction enthalpy, the stronger will be this tendency.

None of the above equilibrium reactions will lead to complete conversion; rather, the actual conversion rate and gas composition will be determined by a state of equilibrium that is more or less close to the theoretical one. This state can be determined by applying *Guldberg* and *Waage's* law [1.4]. As over the temperature range which is relevant for coal gasification the carbon occurs only in solid form and its partial pressure is constant, it may be neglected in the application of this law. Hence, only the partial pressures of the gaseous

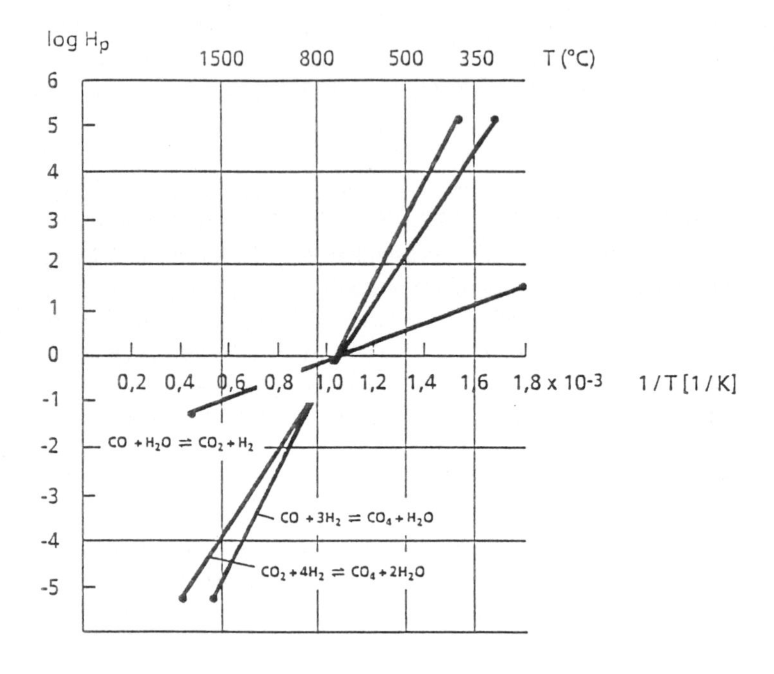

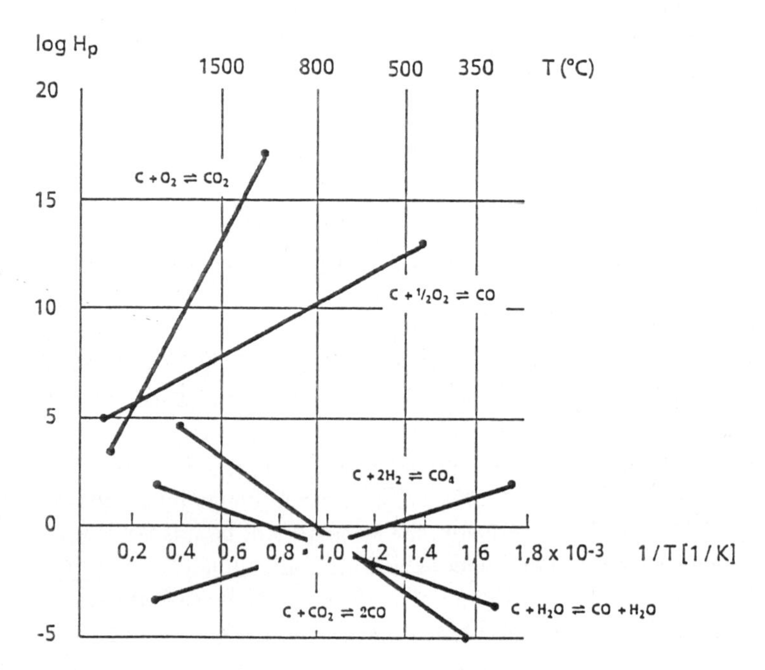

Fig. 1.3. Equilibrium constants of heterogeneous and homogeneous reactions

components need to be considered in heterogeneous gas/solid reactions. Strictly speaking, the fugacities rather than the partial pressures of the various gaseous components would have to be used at elevated pressures, but for almost all practical purposes, the results obtained with the partial pressures will be accurate enough.

1.2.4.3 Kinetics

As mentioned already in the previous chapter, time does not play a role in the formulation of equilibrium conditions. The mass and energy balances and the maximum conversion rate that can be achieved when the system is in thermal equilibrium are therefore not sufficient to define the dimensions of the best suited coal gasification equipment, which will here be termed *the gas generator*. A knowledge of how the gasification reactions proceed with time is therefore indispensable. To measure and, wherever possible, calculate this dependence on time is the object of reaction kinetics. Dealing with details of kinetic laws would go beyond the scope and purpose of this book and the reader should turn to the relevant literature [1.4].

Generally it may be said that kinetics is essentially determined by the coal heating rate in the gasifier and by the reactivity of the coal. Looking at combustion of some of the coal simply as a way of supplying heat – as it might well be introduced also by heat carriers from outside – and neglecting the chemical reactions involved, it will be noticed that carbon gasification is preceded by pyrolysis, which itself leads to the formation of carbon and intermediate products. In a very simplified form, the two processes may be described as follows:

$$CH_xO_y \longrightarrow C,\ \text{tar, higher hydrocarbons,}\quad CH_4, CO, CO_2, H_2O$$

$$C,\ \text{tar, higher HC} + H_2O \longrightarrow CO + H_2$$

Small heating rates as they are characteristic for countercurrent systems in fixed-bed reactors will cause coal pyrolysis, and hence tar formation, to begin at about 350 °C. Since at this temperature the reaction of tar and carbon with steam proceeds very slowly, the concentration of intermediate products such as tar or other hydrocarbons which are liquid under normal conditions is very high. Owing to the countercurrent principle, these products are discharged from the reactor together with the gasification gas itself.

If, on the other hand, the coal is heated very rapidly – as it is the case in fluidized bed and entrained bed systems – pyrolysis is shifted towards much higher temperatures. At such temperatures, the intermediate products and carbon are so quickly reacted with steam and the entire gasification process is accompanied by such strong remixing effects that the product gas does not contain any tar.

Coal reactivity, which is the second factor with a significant influence on kinetics, depends on

- the porosity of the coal, i.e. the inner surface and active centres
- the crystal structure of the carbon
- the catalytic effects of the minerals in the coal.

Young coals with a low degree of coalification such as brown coal or young bituminous coal have a large specific surface and consequently good reactivity. Old coals, anthracite in particular, have a very poor reactivity. Reactivity is enhanced above all by alkalines, especially potassium compounds. Although the mechanism leading to this effect is not yet absolutely clear, it is generally assumed that these substances act as catalysts in the reaction of oxygen with carbon.

Other factors influencing the reaction velocity, in addition to the heating rate and the reactivity of the coal, are its exterior surface and the reaction temperature. An increase in the exterior surface, as it is achieved in the gasification of coal dust, increases the reaction velocity in the same way as a higher temperature. At low temperatures, the reaction velocity is determined by chemical conversion, which is much slower than diffusion between the gasifying agent and the inner coal surface ($\eta \ll 1$). A higher temperature will increase chemical conversion so that the concentration of the gasifying agent in the coal particles decreases and the diffusion rate within the particle determines how the reaction proceeds ($\eta<1$). If the temperature is further increased, the chemical reaction will take place so quickly that the gasifying agent is completely reacted already on the particle surface and diffusion into the pores comes to an end ($\eta=1$). At this high temperature level, the reaction velocity is determined exclusively by diffusion through a boundary layer surrounding the coal particle. The reaction is accelerated also by a higher gas flow rate relative to the solids and, within certain limits, by an increase in gasification pressure. In the lower temperature range, the rise in reaction velocity is almost linear over a pressure range up to approximately 100 bar, while at high temperatures an increase in pressure does not lead to more than a parabolic increase in the reaction velocity, and even that only up to pressures around 10 bar.

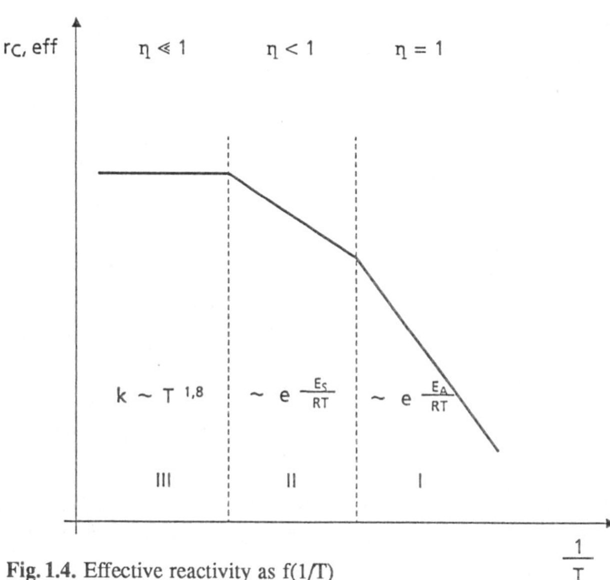

Fig. 1.4. Effective reactivity as f(1/T)

If the effects of the chemical reaction and diffusion are combined into a coefficient r_{eff} for the effective reactivity, the result can be illustrated by an *Arrhenius* diagram (Fig. 1.4) for the three above-mentioned temperature ranges.

1.3 Technology of Gasification Processes

It has been mentioned in the previous chapter that kinetic behaviour can to a certain degree be calculated in advance. Together with the results of laboratory tests, these calculations make it possible to select the most suitable gasification process for a particular type of coal. Reactivity above all plays a major role in this selection. However, unless empirical data are available for the chosen type of coal, tests – at least on a semi-industrial scale – are indispensable to define the optimum process parameters and dimensions of the gas generator. Only semi-industrial or industrial-size equipment will allow those processes which are relevant for gas generation from the compressed mix of organic and mineral components that has been termed coal to be studied with sufficient accuracy. In addition to the grain size referred to above, these tests can be used to derive data on the *mechanical* behaviour of the coal and its ash, i.e. on its caking and burning behaviour, the stability of the coal lumps during heating, the ash melting and solidification behaviour etc. Such large-scale tests will also provide more accurate answers than a laboratory test about the formation of byproducts from the other substances accompanying the coal. These answers are very important especially for the design of the downstream gas treatment facilities. If they are disregarded because of their low concentrations, the byproducts may not only disrupt and damage the equipment but also lead to considerable problems in conforming to environmental legislation.

As illustrated in Table 1.2, only a limited number of industrially proven (and still up-to-date) processes can be used for methanol syngas production. For all practical purposes, it is only autothermal gasification with oxygen and either with or without steam that stands for selection. Processes of this type are available for all three gasifier systems, i.e. fixed-bed, fluidized-bed and entrained-bed gasification systems. The most appropriate characteristic to distinguish between the three gas/solids reaction systems is to look what state of motion the solid grains are in. If the coal lumps are moved either by gravity or mechanically, but not by the gasifying agent or reaction gas, the reactor is termed a fixed-bed one, or – to account for the slow but continuous movement of the coal during gasification – a moving bed gas generator. This mode of operation is typical for the LURGI Pressure Gasification Process.

If the bed is loosened as the gas flow rate increases and becomes turbulent when the minimum fluidizing velocity is exceeded, the system is called fluidized-bed gasification. This principle is used in *Winkler* gasifiers.

If the gas velocity accelerates the coal particles beyond mere suspension so that they are pneumatically lifted, gasification takes place in a cloud of dust and the system is termed entrained-bed gasification. This principle is today used by

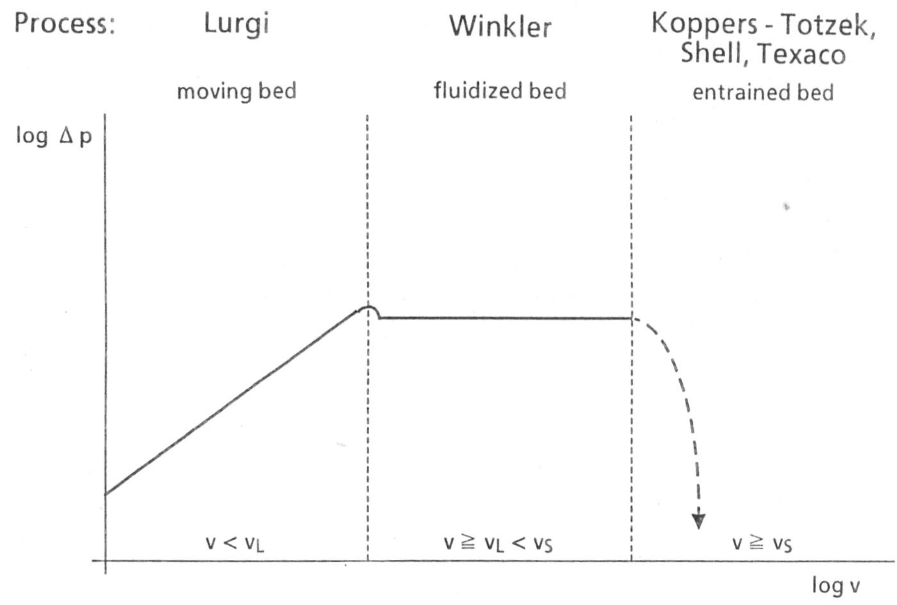

Fig. 1.5. Dependence of pressure drop upon gas velocity

the *Koppers-Totzek*, TEXACO slurry gasification, and the SHELL Coal Gasification Processes.

Figure 1.5 shows pressure drop as a function of gas velocity for the three different gasification systems [1.5]. In this figure, v_1 stands for the minimum fluidizing velocity, v_s for the stable fluidizing velocity. A moving bed can be used to gasify coal particles with sizes between 4 and 30 mm, and in individual cases, where the stability of the coal permits doing so, lumps of "diameters" as high as 80 mm are used. Fluidized-bed technology normally requires particle sizes between 0.1 and 10 mm, whereas entrained-bed gasifiers call for coals ground down to 0.1 mm or less. Moving-bed reactors operate in temperature ranges between 800 and 1 100 °C, fluidized-bed reactors use temperatures of 780–840 °C, and entrained-bed systems operate at temperatures around 1 500 °C. Throughput depends on the types of coal used; for gas flame coal, for instance, the throughput per m^3 shaft volume reaches about 600 kg/h in a moving bed (dry bottom), 1 200 kg/h in a slagger, approximately 500–700 kg/h in a fluidized bed at pressures around 40 bar, and about 2 000 kg/h – also at 40 bar – in an entrained bed.

1.3.1 Moving-Bed Gasification

The commercially interesting field of coal gasification had its origins in the gasification of coarse-grained coal in a quasi-stationary bed. With this system, coal is fed to the top of the gasifier shaft and the gas is also withdrawn overhead. The gas flow rate in the shaft is smaller than the minimum fluidizing velocity. The gasifying agents – steam and oxygen – are blown into the bed from below, i.e. in a countercurrent to the fuel movement. The ash may be discharged either in solid form (dry-bottom gasifiers) or as a liquid phase (slagging gasifiers). The former group includes not only the atmospheric predecessors of the *LURGI pressure gasifier*, which are still rarely found today and whose conversion from air to oxygen operation has never won industrial favour [1.6] but above all the LURGI pressure gasifier itself. Some 80 million cubic meters of gas per day are today produced with this type of generator throughout the world. Similarly, the old type of pressureless slagging gasifier that was developed by *Leuna* and BASF and successfully operated for many years with oxygen and steam at only slightly more than atmospheric pressure, has meanwhile disappeared from virtually all production facilities. An advanced version of the dry-bottom gasifier is the slagging gasifier.

LURGI pressure gasification, the method by which today more gas is produced from coal than by all the other methods together, is the only process to gasify lumpy coal with oxygen and steam that has gained worldwide recognition. Suggestions made by *Drawe* at the Technical University in Berlin were taken up by LURGI (*Danulat* and *Hubmann*) and developed into an operative process in the thirties. The process can be used on virtually all the known types of coal, whether they contain only 2 % ash (Australian brown coals) or more than 40 % (South African bituminous coal), whether they are young and highly reactive bituminous coals or anthracites, and irrespective of the question if they are caking or non-caking. A pre-requisite for successful gasification and smooth operation is a particle size of not less than 3 mm. Coals of smaller size have to be briquetted before they can be gasified. Although the process has been successfully used at pressures up to approximately 100 bar, pressures between 25 and 35 bar have been found most appropriate for the production of methanol syngas as higher pressures would unreasonably increase the methane content of the gas.

Figure 1.6 shows a section through a LURGI pressure gasifier of the latest generation (dry-bottom gasifier). A coal lock above the gasification space receives the feed coal discontinuously from the storage bin and releases it continuously into the gasifier where it is first loaded onto a distribution plate to be evenly distributed over the shaft cross section. If heavily caking coals are to be gasified, the coal distribution facility is equipped with scraper arms to prevent the formation of a cohesive cake. The coal then gets into the gasifier shaft where it is gasified in a quasi-stationary bed (in fact it sinks down at a very low speed) in a coutercurrent to the steam and oxygen serving as gasifying agents.

These gasifying agents enter the gasifier below the ash grate whose design provides for an almost even distribution of them across the entire gasifier cross section. Flow controllers in the oxygen and steam feeder lines ensure that the coal bed subsides evenly in proportion to the flow of gasifying agents. Hence, no complex coal metering system is required – the coal flow looks after itself [1.7]. Within the gasifier, the coal moves through several zones blending into each other

- drying,
- low-temperature carbonisation

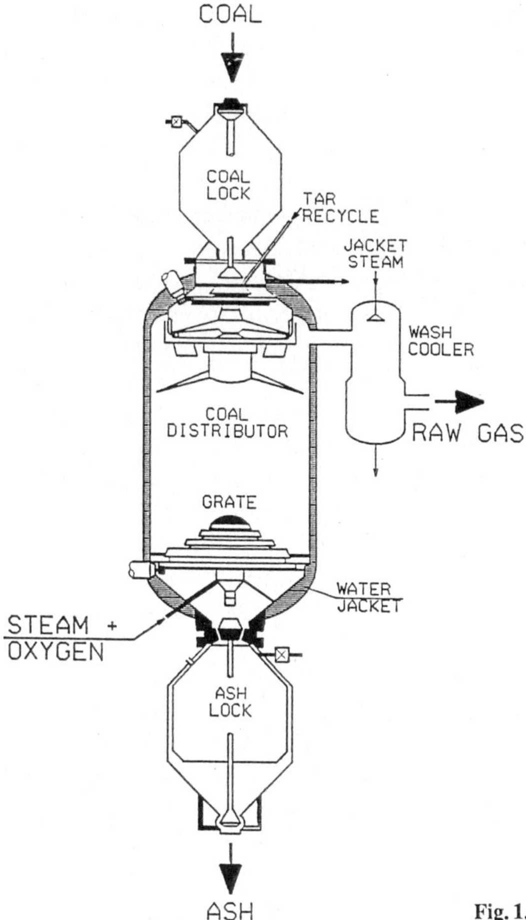

Fig. 1.6. *LURGI* dry bottom gasifier

- gasification
- combustion.

The ash is almost completely burnt out; as a rule, it contains no more than 1 wt. % of the carbon that had originally entered the system. A rotating grate in the gasifier bottom discharges the ash continuously to the ash lock; it has already been precooled to a considerable degree (300–400 °C) by the gasifying agents flowing upward through the grate and can be discharged intermittently into a quench bath from which suitable conveyor systems deliver it to the tip.

The raw gas leaves the gasifier overhead and enters a quench cooler, termed the *wash cooler*, to be cooled by means of gas liquor from the condensation stage. It absorbs a certain amount of steam and leaves the wash cooler at a temperature of about 180–190°C depending upon the prevailing gasification pressure. As the gas is cooled down, most of the high-boiling hydrocarbons are condensed and the dust carried over from the gasifier is washed out. As shown in Fig. 1.7, the gas liquor leaving the wash cooler is combined with the condensate from the

Coal pressure gasification equipped with 36 dry bottom gasifiers

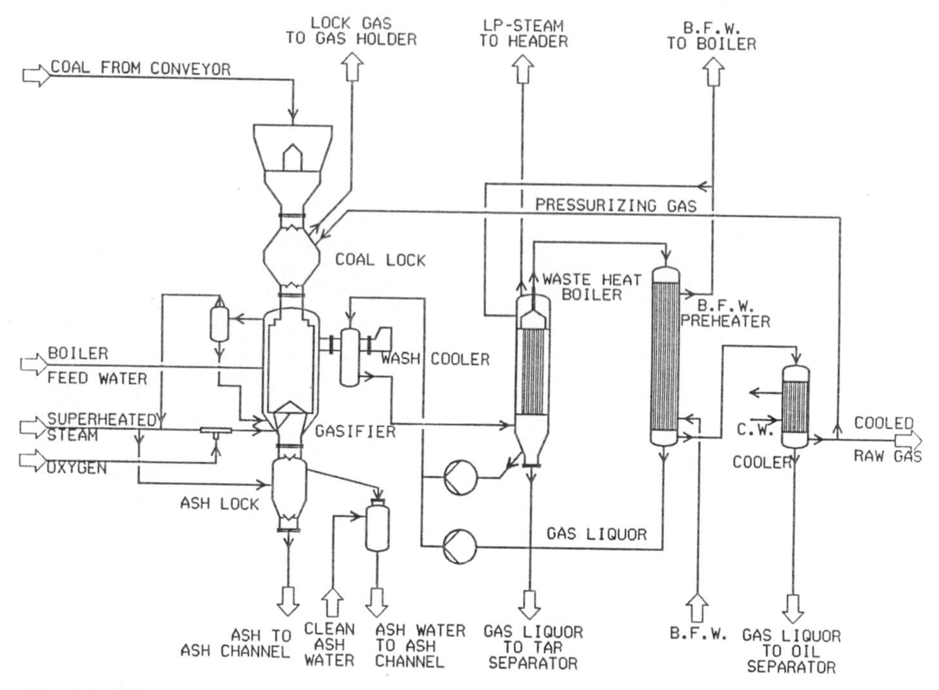

Fig. 1.7. *LURGI* coal gasification process

downstream cooling stages and transferred to the condensate separation unit in which a dust-laden highly viscous tar is withdrawn. This tar is recycled to the gasifier not only in order to recycle its carbon content but also to serve as a *dust catcher* on the surface of the coal bed.

At regular intervals, the gas outlet is mechanically cleaned by a scraper on-line to prevent fouling by caking solids. All coal feeding, ash discharge and outlet cleaning operations are fully automated. They are controlled by hydraulic systems whose operating cycles are appropriately adjusted to the properties of the coals to be handled.

The reaction space itself is enclosed by an inner shell with relatively thin walls which is accommodated inside the thick-walled pressure vessel. The space between the inner and outer shells is filled with boiling water to cool the inside. This boiling water is pressurized to a level slightly higher than the reactor pressure; it is used to generate steam which can be added to the gasifying agent.

Figure 1.8 illustrates the temperature curves for the four reaction zones within the gasifier together with the respective gas compositions. It is clearly recognizable that CO and CO_2 have practically reached their final concentrations already in the area between the combustion and the gasification zone, whereas the methane content rises steeply at the transition from the gasification to the carbonisation zone and the hydrogen content increases more in the lower and less in the upper half of the reactor. Table 1.4 gives an overview over the raw gas analyses that may result from different types of feed coals [1.6]. It can be seen

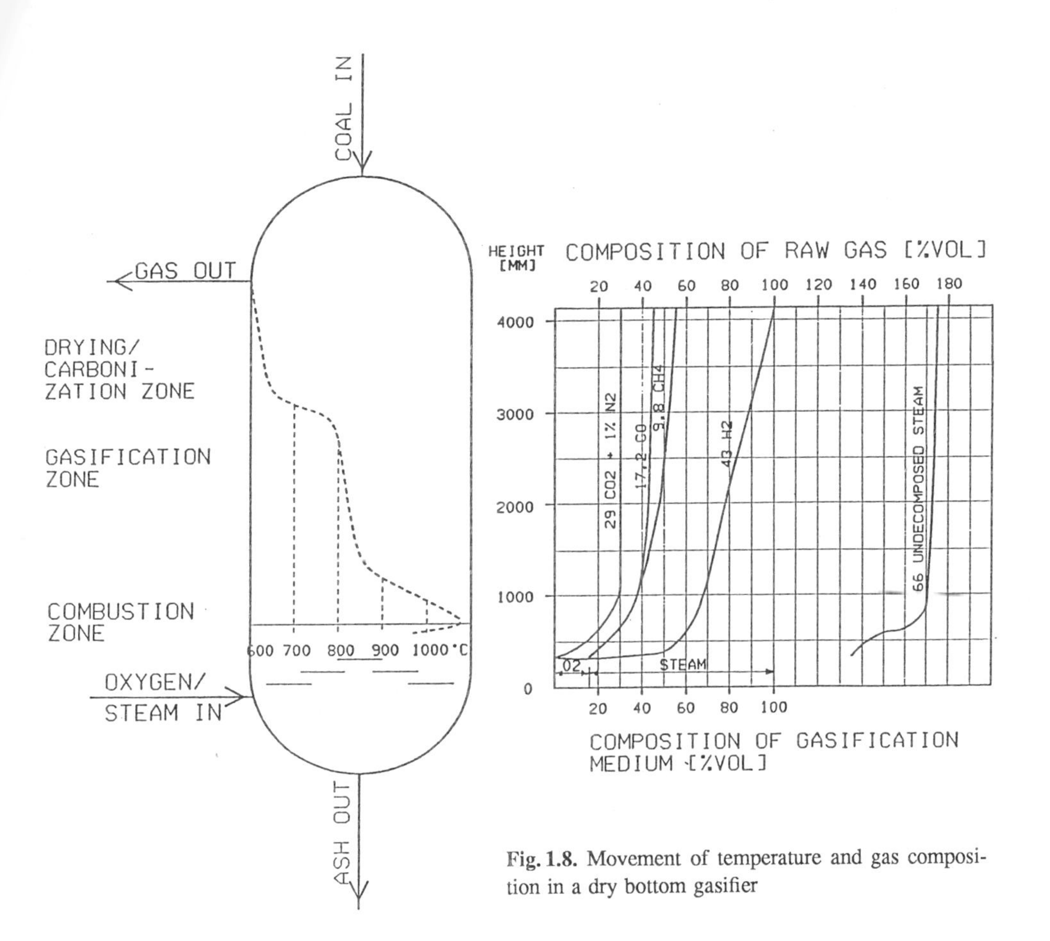

Fig. 1.8. Movement of temperature and gas composition in a dry bottom gasifier

that the methane content decreases with increasing coalification. All figures are derived from gasification at about 20 bar [1.6]. The efficiency of the LURGI pressure gasification system, expressed in terms of HHV(Gas+Liquids)/HHVCoal lies around 90 %.

The principle diagram of the LURGI pressure gasification process is shown in Fig. 1.7. Having been graded to the proper particle size, the coal is stored in bins above the coal lock and from there slides down into it in accordance with the scheduled lock cycles. While it is being filled, the lock is open at the top and closed at the bottom. As soon as the lock is full, the top closes and the lock is pressurized with cold raw gas to the required gasifier pressure. When this pressure has been reached, the bottom of the coal lock opens and the coal drops onto the distribution plate to be fed to the gasifier as described above. When the coal lock is empty, the bottom closes and the lock is depressurized. The resulting gas, termed *lock gas*, is collected in a gasholder and may either be used for firing purposes or recycled to the raw gas flow.

Ash discharge through a lock has already been mentioned before. The procedure is analogous to coal feeding. The ash lock is pressurized by means of

Table 1.4. Results of gasification of coal by *LURGI* pressure gasification at 20 bar

Type of Coal:	Lignite	Bituminous	Anthracite
Coal Components (m.a.f.)			
C % wt.	9.50	77.3	92.1
H	4.87	5.9	2.6
S	0.43	4.3	3.9
N	0.75	1.4	0.3
O	24.45	11.1	1.1
Raw Gas Composition (dry)			
$CO_2 + H_2S$ % vol.	30.4	32.20	30.9
CnHm	0.4	0.79	0.4
CO	19.7	15.18	22.1
H_2	37.2	42.15	41.0
CH_4	11.8	8.64	5.6
N_2	0.5	0.68	0.8
m^3 $(CO + H_2)$/ 1000 kg coal (m.a.f)	1050	1340	1480
kg steam/ m^3 $(CO + H_2)$	1.18*	1.93*	1.34*
m^3 oxygen/ m^3 $(CO + H_2)$	0.17	0.28	0.30

*including jacket steam

superheated steam which is then condensed by quenching with water in a separate vessel when the lock is depressurized again.

As briefly described already, the raw gas leaves the gasifier overhead and enters a quench cooler, in which it is quenched with gas liquor from the bottom of the waste heat boiler (see Fig. 1.7) whereby high-boiling hydrocarbons are condensed and some of the gas liquor is evaporated so that the steam dew point of the raw gas is raised. Thereafter the raw gas goes to the waste heat boiler. Because some of the high-boiling hydrocarbons are still not condensed at this stage, the waste heat boiler is designed for a certain self-cleaning effect to keep the heat exchanger tubes free from permanent fouling. The raw gas flows through the tubes from the bottom upward, while the condensate runs downward, is collected in the bottom, and is then partly transferred to the gas liquor separator and partly used as quench water in the wash cooler.

According to the diagram shown here, the major part of the remaining heat in the raw gas is used in a downstream heat exchanger to heat up boiler feed water before the raw gas flows through a trim cooler to be cooled to ambient temperature. Needless to say, other ways of utilizing the heat are conceivable from case to case.

Table 1.5. Gasification of Kentucky bituminous coal by the *LURGI* dry bottom gasifier

Complete Gas Analysis

CO_2 + S * [% vol.]	31.50	* S detailed:	
CO	16.43		
H_2	41.24	COS [v/ppm]	180
CH_4	8.97	H_2S	15 300
C_2H_4	0.11	Mercapt. S	600
C_2H_6	0.45	Tiophenes	5
C_3H_8	0.07	CS_2	100
C_4H_{18}	0.10		
C_4H_{10}	0.05	Others:	
C_5H_{10}+	0.18	HCN [v/ppm]	22
N_2	0.59	NH_3	39
Ar	0.36	NO_x	.02

Traces of: As, Ca, Cd, Cl, Fe, Hg, Na, Mo, Zn
Organic acids, Phenols

If not used as quench water, the tar and oil-laden gas liquor from the waste heat boiler is split up in the tar separator (see Sect. 5.1) into a dusty tar, a thin tar, and a gas liquor fraction. The condensate from the feed water preheater, like some of the condensate from the waste heat boiler, is used as quench water in the wash cooler. The trim cooler condensate, which essentially contains only hydrocarbons with boiling points of less than 100 °C, is sent to the oil separator to be separated into an oil fraction and a gas liquor fraction. The gas liquor passes an additional oil trap and then goes to the Phenosolvan unit.

Having been cooled down to ambient temperature, the raw gas enters the gas cleaning section. As it still contains non-negligible quantities of condensable hydrocarbons, which would disrupt most gas cleaning processes either by foaming or by detracting from the cleaning efficiency through hydrocarbon buildup in the wash liquor, the gas cleaning section has to be preceded by a number of precleaning stages in which virtually all the condensable hydrocarbons are removed from the gas (see Sect. 2.3). How intricate in spite of this pretreatment gas cleaning to synthesis purity may still be, can be seen from the complete raw gas analysis after the precleaning stages shown in Table 1.5.

For start-up, the gasifier is filled with coal which is then heated to ignition temperature (approximately 450 °C) by means of superheated steam. As soon as this temperature has been reached, ignition is initiated by admitting air. When it has been established that the bed *burns* evenly – checked by the oxygen content of the raw gas – the system is gradually transferred from air to oxygen and steam. Whenever the gasifier is not required to produce but is only to be kept operative, i.e. *warm*, for a quick restart, this condition may be achieved by admitting a small flow of air.

Industrial versions of the LURGI dry-bottom gasifier operate at pressures around 30 bar and with nominal shaft diameters of as much as 5 m. The maxi-

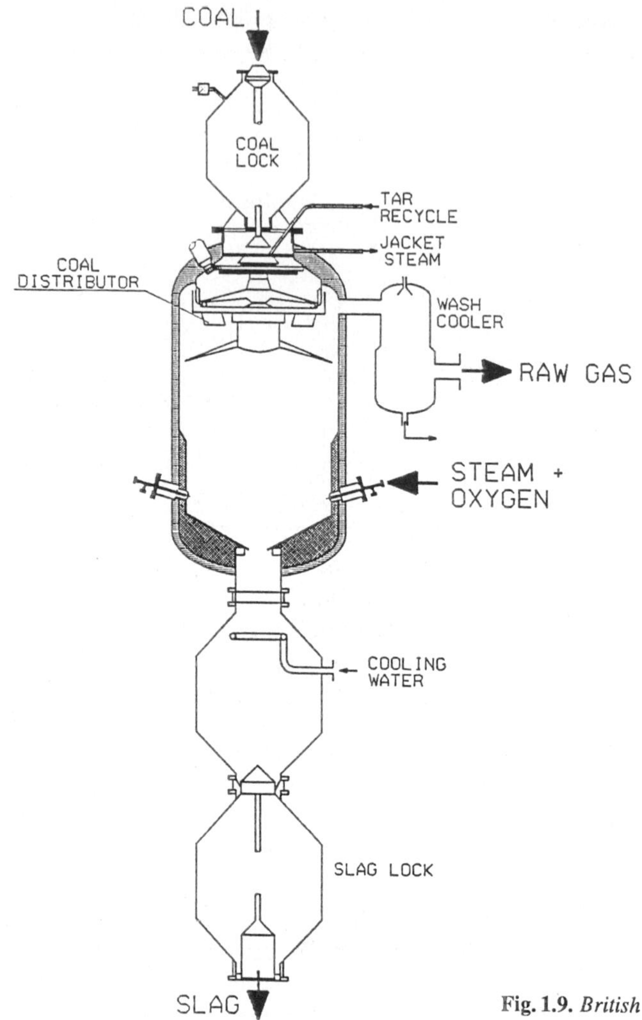

Fig. 1.9. *British Gas/LURGI* slagging gasifier

mum coal throughput for which there is not yet any risk of dust discharge from the gasifier is about 45 t/h (m.a.f.). Allowing for the reforming of the liquid hydrocarbons and phenoles and for secondary methane reforming (see Sect. 2.4.2), this is sufficient to meet the syngas demand of a 900 tpd methanol plant.

The *British Gas/LURGI* slagging gasifier is a further development of the LURGI dry-bottom gasifier which was begun by *British Gas Corporation* and matured to industrial application together with LURGI. As illustrated by Fig. 1.9, the design details of the slagger top are largely the same as for the LURGI pressure gasifier. However, the ash leaves the latter in solid form whereas from the slagger it is discharged as a liquid. The engineers who developed the slagger had three main objectives in mind:

– increasing the CO and H_2 yields,

- suitability for coals having a very low ash melting point or improving the cost-effectiveness when using such coals,
- solving the fines problem.

A comparison with the features of the LURGI pressure gasifier makes it clear that these three are the only aspects that impose certain limitations on its universality and economic merits, especially for the production of syngas. For the slagger, on the other hand, one has to accept certain restrictions with respect to the reactivity, ash content and ash melting point of the coals to be gasified. It is particularly well suited for the gasification of low-reactivity coals such as US Mid-Western and Eastern types, UK coals, coals from the German Ruhr area, as well as petroleum coke. The ash content of the feed coals should not exceed 20 wt. % and their ash melting point should be less than 1 200 °C.

Table 1.6 highlights the differences in the operating principle between the slagger and the dry-bottom gasifier [1.9]. Whereas the CO + H_2 yield of a dry-bottom gasifier is 1 340 m^3 per 1 000 kg of coal (m.a.f.), it reaches 1 920 m^3/1 000 kg for the slagger. This remarkable difference results for a small part from the decrease in methane content due to the higher gasification temperature in the slagger, but for the most part is caused by the shift in the water gas equilibrium towards CO as the steam feeding rate is less than one sixth that of the dry-bottom gasifier. Admittedly, when the CO_2 content is reduced to about 3 %, the stoichiometric ratio (H_2 – CO_2)/(CO + CO_2) in the dry-bottom gasifier has the level of 2.1 which is desirable for methanol synthesis (see Sect. 3.2) whereas the slagger gas has too high a CO content. However, the surplus CO can be converted to H_2 and CO_2 in a catalytic stage requiring much less steam than a gas generator in which the water equilibrium *freezes* at about 900 °C for lack of catalytic action.

In comparison with the raw gas from the dry-bottom gasifier, the slagger gas has still other advantages for methanol production which should not be ignored:

- It contains so little CO_2 that non-selective desulfurization (see Sect. 2.3.1) produces a *Claus* gas which can be treated in a standard *Claus* unit without previously requiring enrichment.
- The lower CH_4 content of the slagger gas results in a smaller specific syngas demand per ton of methanol and thus saves compressor energy and reduces the cost of the synthesis loop and the catalyst.
- The gas condensate rate is only about 1/5 that of the dry-bottom gasifier.

The so-called cold gas efficiency of a dry-bottom gasifier, expressed as the ratio of heat output with the cold gas to heat input with the coal, HHVGas/HHVCoal both in terms of their higher heating value, reaches about 80 %. If one allows for the hydrocarbon liquids obtained as the gas is cooled, the gasification efficiency increases significantly. Taking account of the syngas yield from these liquid hydrocarbons and from the phenols, which may be gasified e.g. by the partial oxidation process (Sect. 2.4), the cold gas efficiency of a dry-bottom gasifier is approximately 90 %. Compared to this the corresponding figures for the slagger are 88 and 90 % respectively.

Table 1.6. Results of gasification of Illinois no. 6 bituminous coal by *LURGI* dry bottom and *British Gas/LURGI* gasifier

Coal Components (m.a.f.)			
C	% wt.	77.3	
H		5.9	
S		4.3	
N		1.4	
O		11.1	
Raw Gas Composition (dry)		**Dry Bottom Gasifier**	**Slagging Gasifier**
CO_2	% vol.	30.89	3.46
H_2S + COS		1.31	1.31
CnHm		0.79	0.48
CO		15.18	54.96
H_2		42.15	31.54
CH_4		8.64	4.54
N_2		0.68	3.35
NH_3		0.36	0.36
m^3 (CO + H_2) / 1000 kg coal (m.a.f.)		1340	1920
kg steam / m^3 (CO + H_2)		1.93*	0.20
m^3 O_2 / m^3 (CO + H_2)		0.28	0.23
kg naphtha, tar, oil, phenolics / 1000 kg coal		81	19

*including jacket steam

The coal feeding and distribution system for the slagger is the same as for the dry-bottom gasifier, and so are the shell-cooling systems and the gas outlet design (Fig. 1.9). But while the solid ash is discharged from the dry-bottom gasifier by means of a rotary grate, which at the same time ensures an even distribution of steam and oxygen as gasifying agents, the slagger bottom had to be fundamentally redesigned. The gasifying agents are introduced and distributed via a system of tuyeres reaching into the gasifier to a level only slightly above the molten slag bath. These tuyeres can be used also to bring byproducts into the combustion/gasification zone such as tar, oil or naphtha, which are rarely desired in a methanol production plant. Coal fines, which would affect gasification in the shaft if introduced through the coal lock in excessive quantities, can also be blown in at this point. As in the case of the dry-bottom gasifier, controlled feeding of the gasifying agents ensures an even lowering of the coal bed.

The molten ash is collected underneath the combustion zone and is drained through the slag tap into the slag quench chamber where it is quenched with water to solidify instantaneously. The solid ash is then discharged via an ash lock similar to the dry-bottom gasifier system. The slag consists of an inert glassy frit which can be tipped without polluting the environment. It has been classified as non-hazardous by US standards.

Heat recovery from the raw gas and its further cooling to ambient temperature as well as the systems to separate high-boilers, low-boilers and gas liquor are virtually the same for both types of gas generators. The slagger has proven its merits on an industrial scale up to pressures around 25 bar. The largest shaft so far measures 2.5 m in diameter and can handle a max. coal throughput of some 20 t/h; this would be approximately equivalent to the syngas demand of a 500 tpd methanol plant. Hence, the output of a slagger in terms of its shaft cross section is about twice as high as that of a dry-bottom gasifier. The start-up and hot stand-by procedure for the slagger is the same as for the dry-bottom gasifier.

1.3.2 Fluidized Bed Gasification

Both coal gasification in a fluidized bed and the fluidized bed technology itself date back to *F. Winkler*, who lent his name to the atmospheric *Winkler Gasification Process* of the twenties and to the *High-Temperature Winkler (HTW)* developed during the fifties. F. *Winkler* had observed that fine-grained, loosely poured material, when gas is blown through it at a certain minimum velocity, is loosened to a degree where it assumes the physical properties of a liquid [1.8].

The gasification principle remains the same for both process variants. In a first reaction zone, fine-grained coal, suspended by the gasifying agent, is pregasified and partially oxidized. In a second reaction zone, the coal particles leaving the first zone then undergo an exothermic reaction with oxygen.

The use of the *Winkler* gasification process and of the HTW is restricted to reactive, non-caking lignites which have to be dried to less than 8 % moisture content before they can be gasified with oxygen and which should suitably be preheated before entering the gas generator to improve efficiency. Since the ash is discharged from the Winkler process as a solid material, the temperature at the top of the gasifier needs to be kept within an extremely narrow range to ensure that the ash melting point is not exceeded – this would seriously disrupt the fluidized bed and lead to incrustation on the downstream cooling surfaces – while, on the other hand, the temperature has to be kept high enough for satisfactory coal burn-out which is crucial for the cost-effectiveness of the process.

Figure 1.10 is a schematic illustration of the HTW gasifier, while Fig. 1.11 shows the overall arrangement of a Winkler gasification unit [1.9,10]. Both figures will hereafter be used to explain the process in greater detail.

Coal of up to 8 mm grain size is withdrawn from a bin, screw-fed to a coal lock and continuously introduced through the side of the gasifier shaft. In the case of the HTW, the screw feeders deliver the feed pneumatically to the coal locks, which are then pressurized to the gas generator pressure. Steam and

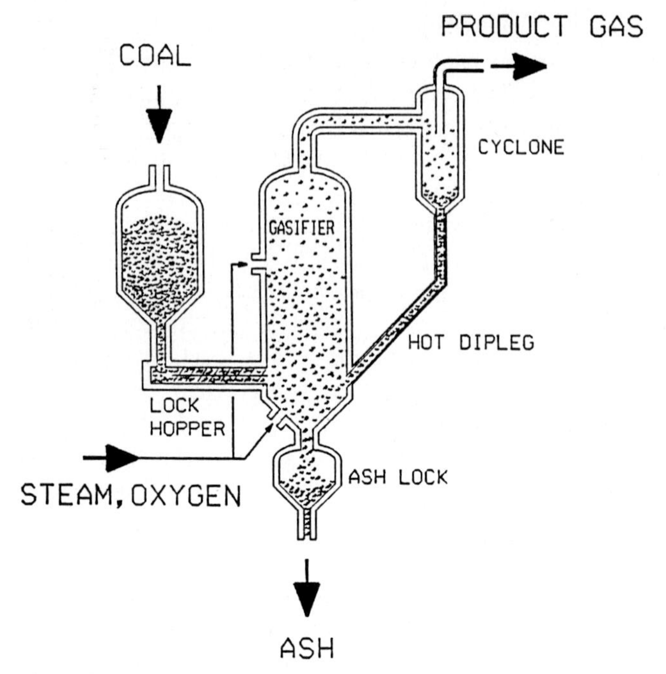

Fig. 1.10. High temperature *Winkler* gasifier (HTW)

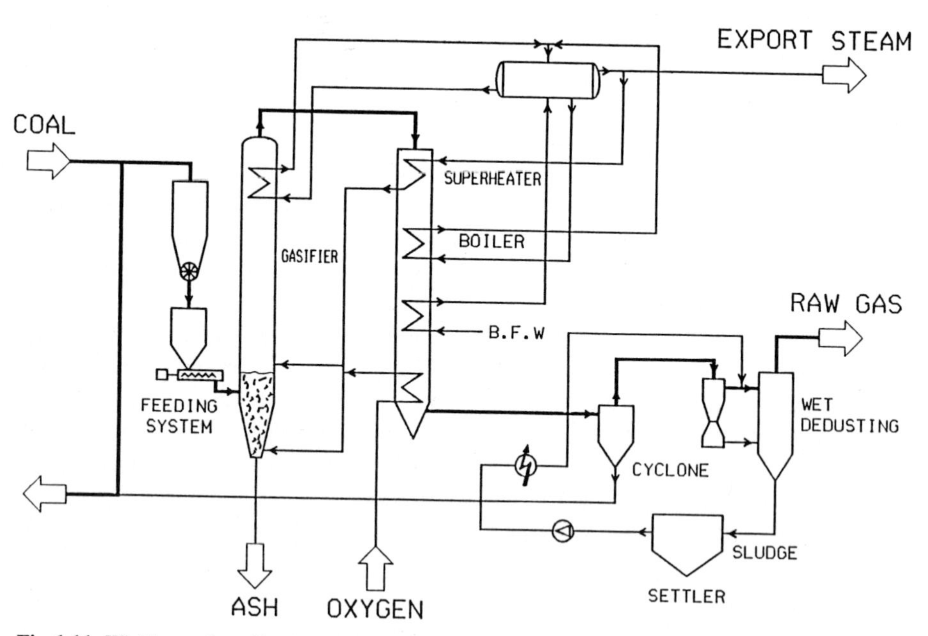

Fig. 1.11. *Winkler* coal gasification process

oxygen are fed to the gasifier shaft at different points. Steam without oxidizing agent is introduced at the bottom in order to keep the fluidized bed of ash and coal particles moving "like a boiling liquid" [1.11] and to cool even larger ash particles sufficiently so that they can be discharged without problems at the gasifier bottom. The gasifier shaft is tapered at the bottom to provide a transition to the ash lock. The fluidized bed itself measures only 2–2.5 m in height, whereas the overall height of the gasifier is more than 20 m. Hence, the secondary reaction zone requires a lot of space. The coal and the gasifying agent react in the shaft bottom at temperatures between 850 and 950 °C. This reaction is a very rapid one. Fluidization in the gasifier bottom produces additional fines which are transported upward into the secondary reaction zone together with the already introduced coal fines. Since additional gasifying agent is introduced at this level, gasification continues above the fluidized bed. For the same reasons as mentioned before, the temperature has to be closely controlled in this secondary reaction zone, too. 1 200° C are considered the upper limit for coals with high-melting ash. Fluidization in the gasifier shaft leads also to classification of the ash and remaining coal particles according to specific mass.

Figure 1.12 shows how ash and coal particles of different sizes are distributed across the height of the fluidized bed [1.8]. The heavier particles – essentially coarse ash particles – move downward and accumulate in an ash lock from which they are discharged by means of screw conveyors. The lighter ash

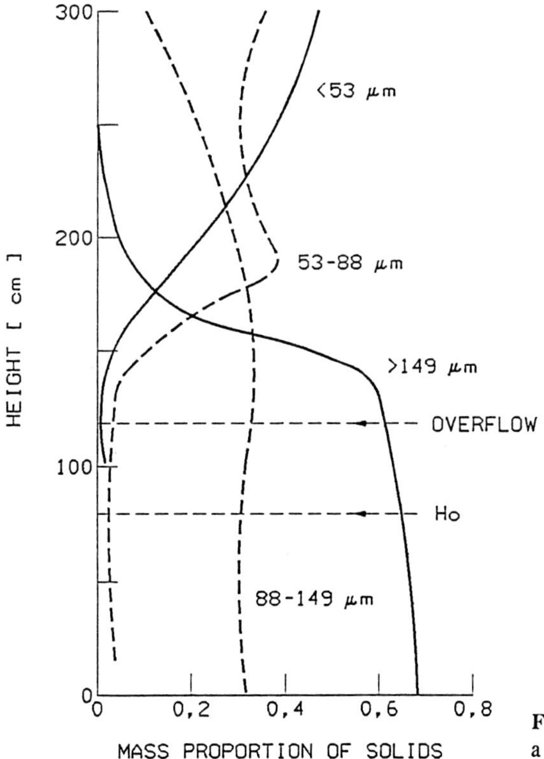

Fig. 1.12. Distribution of particle size in a fluidized bed

particles along with ungasified or incompletely gasified coal fines leave the gasifier at the top and are separated from the gas in a hot cyclone and then returned to the shaft bottom by way of a hot dipleg. Very fine ash particles are finally eliminated from the product gas in a second cyclone and discharged via a separate ash lock. The coarse ash from the first ash lock still contains some 20 % carbon, i.e. more than 4 % of the carbon input, so that it may be added, e.g., to the feed coal for a coal-fired boiler (power station).

In order to prevent sintering ash particles from fouling the top of the gasifier shaft and the gas outlet piping and downstream cooling surfaces, the top of the *Winkler* gasifier is equipped with a radiant boiler at the walls of which the entrained fly ash is cooled by approximately 200 °C. In this way, the fluidized bed can be safely operated at a temperature that leaves only a slight margin to the ash melting point. The radiant boiler uses already some of the sensible heat of the gas to generate steam. Heat recovery is continued in a waste heat boiler – which may be preceded by a steam superheater – and in heat exchangers to preheat oxygen and boiler feed water. The second cyclone used for mechanical dedusting as mentioned above is normally accommodated in an area where the gas has already been largely cooled down. The last gas cleaning stage is a venturi scrubber in which the gas is cleaned to *machine purity*, i.e. to residual dust contents of less than 2 mg/m^3. As the steam/feed ratio is low in comparison with fixed-bed gasification, a fluidized-bed gasifier produces much less waste water. However, since this waste water contains much the same byproducts, except for high-boiling hydrocarbons, as the waste water from fixed-bed gasification, it is preferably recycled to the gasifier after such low-boilers as, for instance, ammonia have been stripped out.

Table 1.7 contrasts the results from the older *Winkler* gasification system with the HTW [1.9,10,12,13]. The analysis of the German brown coal used in the HTW gasifier is probably equivalent to that listed for the standard *Winkler* process. No detailed data regarding this point nor a corresponding gas analysis for the demonstration plant in West Germany have so far been published. The table is, however, based on two different types of lignite. A summary of the undesired byproducts contained in the gas is given in Table 1.8. Coals whose ashes contain very large quantities of lime are already partly desulfurized in the fluidized bed so that the H_2S content of the raw gas is lower than it would be expected from the sulfur content of the coal. The resulting calcium sulfide remains in the ash. The HTW gasification system reaches a cold gas efficiency of 80–82 %, depending on the type of coal used. If the heat required to dry the coal is included, this efficiency drops to something like 76–78 %.

At present, three classical *Winkler* gasifiers are still operating throughout the world; no new ones have been built for the past 25 years. The only commercial HTW gasifier so far was conceived as a demonstration plant for a coal throughput of approximately 30 t/h and has been operating for several years now. The gas is used to produce methanol. Its coal throughput meets the demand of a 400 tpd methanol plant. Another HTW plant of the same size is currently under construction.

Table 1.7. Results of gasification of lignites by *Winkler* gasification

	Standard Winkler Gasification (atm. pressure)		HTW-Gasification (10 bar)
Coal Type:	West German Lignite	North Dakota Lignite	West German Lignite
Components (m.a.f.)			
C [% wt.]	68.1	72.6	
H	5.2	4.8	
S	0.5	1.3	n.a.
N	1.0	1.0	
O	25.2	20.3	
Ash	5.0	9.0	
Moisture	8.0	12.0	
Raw Gas Composition (dry)			
$CO_2 + H_2S$ [% vol.]	13.8	9.4	23.3
CO	48.2	53.0	40.3
H_2	35.3	33.7	31.8
CH_4	1.8	3.1	3.7
N_2 + Ar	0.9	0.8	0.9
m^3 ($CO + H_2$) / 1000 kg coal (m.a.f.)	1500	1730	1580
kg steam / m^3 ($CO + H_2$)	0.43	0.14	0.25
m^3 oxygen / m^3 ($CO + H_2$)	0.41	0.27	0.21

Table 1.8. By-products from HTW-gasification of lignite [mg/m^3]

CS_2	
COS	20
RSH	
C_2H_2	
C_2H_4	< 20
C_2H_6	
C_6H_6	< 3500
Cl_{total}	< 0.2
NH_3	< 5
CN	15
Phenols	2
Naphthalene	10
Naphthol	10
Traces of Hg, Cd, Ba, Pb, Ni, V, Fe	

1.3.3 Entrained Flow Gasification

In spite of repeated attempts, it took about 60 years until a patent granted already in 1890 [1.13] was successfully developed into an operative plant for the gasification of coal in an entrained flow. The technical basis for the process to function was laid by *H.Koppers* and *F.Totzek*, and the process became known as the *Koppers Totzek Process*. A major contribution to making it operative came from the *US Bureau of Mines* which also sponsored the construction of a first demonstration plant. In addition to the virtually atmospheric *Koppers-Totzek* process – which is described in greater detail below – to the SHELL Coal Gasification Process and to TEXACO's Slurry Process, the entrained-flow technology also spawned the DOW process (Dow Chemical Cie.), as well as the Prenflo (GKT) and *Westinghouse* processes (*Westinghouse Electric Corp.*) the two last of which have so far not yet been used commercially.

A simplified process flow diagram of the atmospheric *Koppers-Totzek* process is shown in Fig. 1.13. The facilities required to prepare (grind down) the coal are not included. Coal that has been ground to dust size and predried if necessary is withdrawn from a service bin and fed to the two or four burners of the gas generator (termed *gasifier heads*) by means of screw feeders. Oxygen (and steam if necessary) is blown into the coal stream in a mixing leg upstream of the burners. The mixture of coal and gasifying agent then enters the reaction zone at high velocity through the burner mouth. The velocity is appropriately selected to prevent backfiring into the mixing leg. The mixture issuing from the burner is so rapidly heated – partly by radiant heat from the wall of the reac-

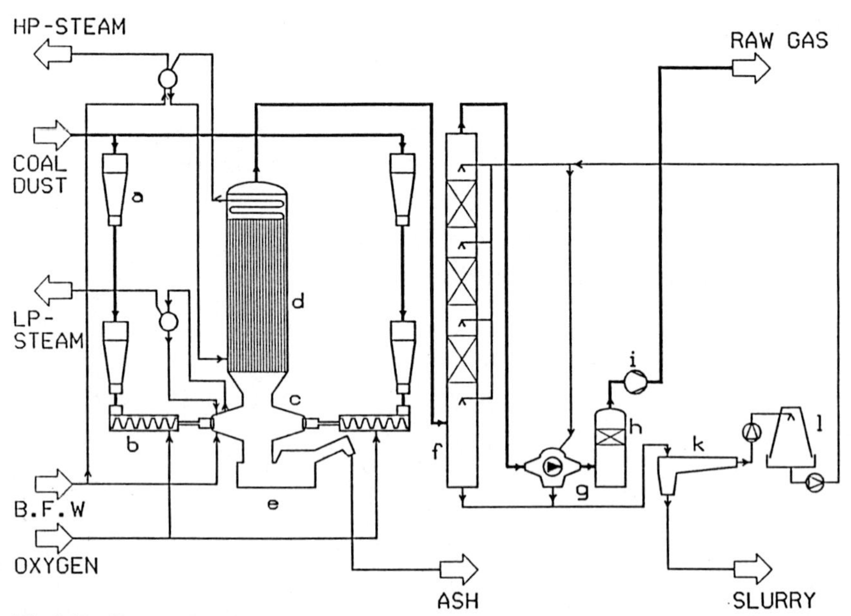

Fig. 1.13. *Koppers-Totzek* coal gasification process. (*a*) Dust bunker system; (*b*) dust screw; (*c*) gasifier; (*d*) waste heat boiler; (*e*) ash extractor; (*f*) wash cooler; (*g*) disintegrator; (*h*) final cooler; (*i*) gas blower; (*k*) settling tank; (*l*) cooling tower

tion chamber and from the flame front, and partly by the erratic hot gas backflow caused by the heavy turbulences – that reaction between coal and gasifying agent sets in spontaneously, lighting a flame. Temperatures at the core of this flame are as high as 2000 °C. Owing to the extremely rapid rise in temperature, the coal particles moving through the flame zone in a cocurrent with the gasifying agent are so quickly gasified that even heavily caking coal does not agglomerate.

The raw gas leaving the reaction space has a temperature of 1500 to 1600 °C so that all devolatilization products and higher hydrocarbons are converted to H_2, CO and CO_2. The residual methane content is down to 0.1 vol. %. Most of the ash leaves the reaction space in liquid form through a central outlet to be quenched and granulated in a water bath underneath the reactor which serves at the same time as a seal to the reaction space. The ash is then discharged from this water bath by means of a scraper belt.

The raw gas leaves the gasifier at the top and enters a waste heat boiler arranged above the gasifier to have most of its sensible heat recovered for the generation of high-pressure steam. After it has been further cooled down in a water cooler, the gas passes a disintegrator and a trim cooler before it is finally delivered via a seal pot to the gas blower to be raised to a higher pressure. The wash cooler and the disintegrator serve to eliminate the fly ash that has been entrained with the gas from the reaction space. This fly ash together with the wash water is discharged to a settling tank to be pumped off as a slurry. The virtually dust-free wash water is sent to a cooling tower and returned to the wash cooler and disintegrator.

A somewhat more detailed impression of the gas generator itself and of the components associated with it can be gained from Fig. 1.14. It shows the

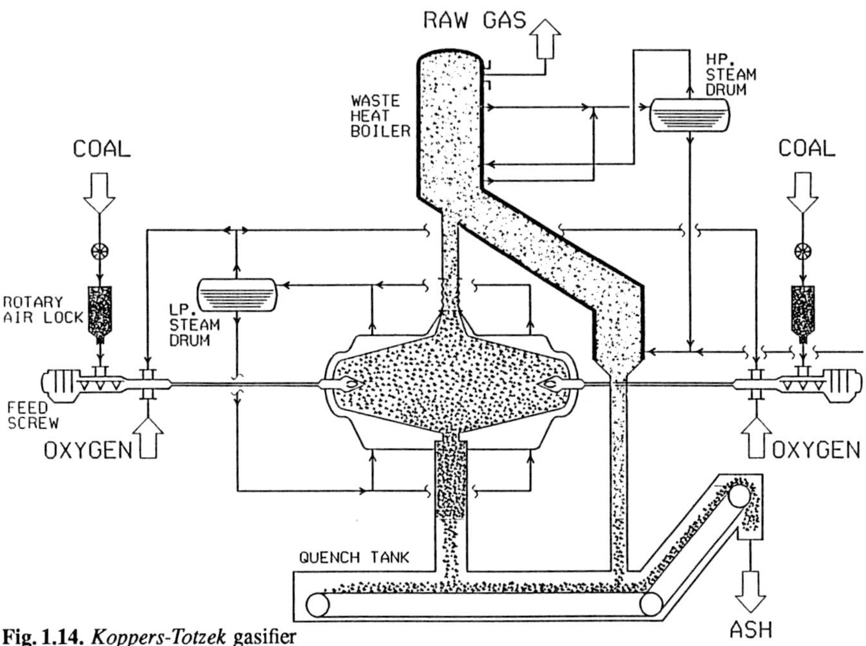

Fig. 1.14. *Koppers-Totzek* gasifier

arrangement of the screw feeders and burners diametrically opposite each other, as well as the service bin with its rotary air lock through which the coal dust is delivered to the feeders. The entire reaction space – like the waste heat boiler on top of it and the downcomer for the liquid ash – is lined with high-temperature refractory material. These sections or parts of them are also enclosed by cooling jackets. The reaction space has a separate low-pressure cooling jacket which recovers the heat dissipated through the bricklining to generate LP steam. The cooling jackets of the waste heat boiler and of the ash downcomer are connected to the HP steam generation loop. It can also be seen from Fig. 1.14 that the ash discharge is isolated from the atmosphere and protected against surge pressures by a relatively deep water seal.

Table 1.9 contrasts the gasification results from three different types of coal – a high-volatile bituminous coal from Illinois, a low-bituminous coal from Zambia, and a lignite from Greece [1.1,9]. With the Illinois coal from Table 1.9, a cold gas efficiency of approximately 67 % can be reached from the dried coal, and approximately 66 % from the moist coal.

Koppers-Totzek gas generators are today still operating in some syngas plants. The largest output reached per gasifier was approximately 50 000 m^3 of raw gas per hour; if lignite is used, this corresponds to the demand of a 450 tpd methanol plant.

Table 1.9. Results of gasification of coal by *Koppers-Totzek* gasification

Coal Type:	Illinois/USA High Volatile Bituminous	Zambia Low Volatile Bituminous	Ptolemais/ Greece Lignite
Coal Components (m.a.f.)			
C [% wt.]	80.1	87.6	66.1
H	5.3	4.6	5.1
S	3.0	1.5	2.2
N	1.6	1.8	1.9
O	10.0	4.5	24.7
Raw Gas Composition (dry)			
CO_2 [% vol.]	8.7	10.2	11.7
CO	62.2	59.4	60.0
H_2	26.8	28.2	26.1
CH_4	traces	races	traces
N_2 + A	1.3	1.8	1.9
H_2S + COS	1.0	0.3	0.3
m^3 (CO + H_2) / 1000 kg coal (m.a.f.)	1750	1890	1455
m^3 oxygen / m^3 (CO + H_2)	0.42	0.43	0.36
kg steam / m^3 (CO + H_2)	0.08	u.a.	u.a.

The *Prenflo Process* (Pressurized Entrained Flow Gasification), an advanced 25–30 bar version of the atmospheric *Koppers-Totzek* process, is currently on its way to becoming operative. A demonstration plant designed for a coal throughput of 2.0 t/h has been operating since 1986. Apart from the fact that the raw gas provided by this process is obtained at an elevated pressure, its improved coal conversion rate promises a markedly better cold gas efficiency. Initially, the new process variant was launched in cooperation between *Koppers* and SHELL, but it is now pursued by the former alone while SHELL went its own way.

This way led SHELL to the development of *SHELL Coal Gasification Process (SCGP)* which is today considerably further advanced than the Prenflo process.

At the beginning of the eighties, the *Royal Dutch SHELL Group of Companies* decided to build a semi-commercial size demonstration plant. Design work on this plant started in 1984 and the plant near Houston,Texas, went onstream around the middle of 1987. It can put through about 10.0 t of bituminous coal or about 17.0 t of lignite per hour [1.14,15]. Although SHELL could rely on almost 30 years of experience with high-temperature gasification of light up to very high-boiling hydrocarbons (methane to propane asphalt), they had in some instances to leave the SHELL *Gasification Process (SGP)* altogether and break entirely new ground for SCGP. With the SGP route, gasification in the reaction space proceeds from top to bottom, while for the SCGP process it takes place from the bottom towards the top. Completely different and considerably more complex feeding systems had to be devised to introduce the dry, unslurried coal, and a remarkable amount of design work had to be performed to cope with the liquid slag and the fly ash. Considering the size of the demonstration plant and the large number of coals gasified in it, the process has to be regarded as operative.

Figure 1.15 shows a simplified process flow diagram for an SCGP plant. Its outline is very similar to that of the *Koppers-Totzek* gasification system and the significant differences become evident only at a closer look. The dust coal, which has to be predried if a raw coal with a high water content is used, enters the pressurized gasification system by way of a coal lock similar to the system described for the LURGI dry-bottom gasifier. Via a service bin, it enters the feed leg to the burners. Unlike the screw feeder system used for *Koppers-Totzek* gasification, the coal dust is transported to the burners by means of nitrogen. Oxygen is added immediately upstream of the burners and with this system, too, the velocity of the coal dust/gasifying agent mixture is kept so high that ignition is not possible before the mix issues from the burner mouth. Reaction in the flame zone is similar to the *Koppers-Totzek* process and the max. temperature reaches approximately 2000 °C as well. The reaction space is a pressure vessel with ceramic lining inside. The reactor wall temperature is controlled by an elaborate jacket-cooling system producing high-pressure steam.

Most of the ash is withdrawn in the form of liquid slag, quenched, and discharged to the outside by way of a system of locks. To prevent entrained ash particles which are still soft from caking the walls of the waste heat boiler

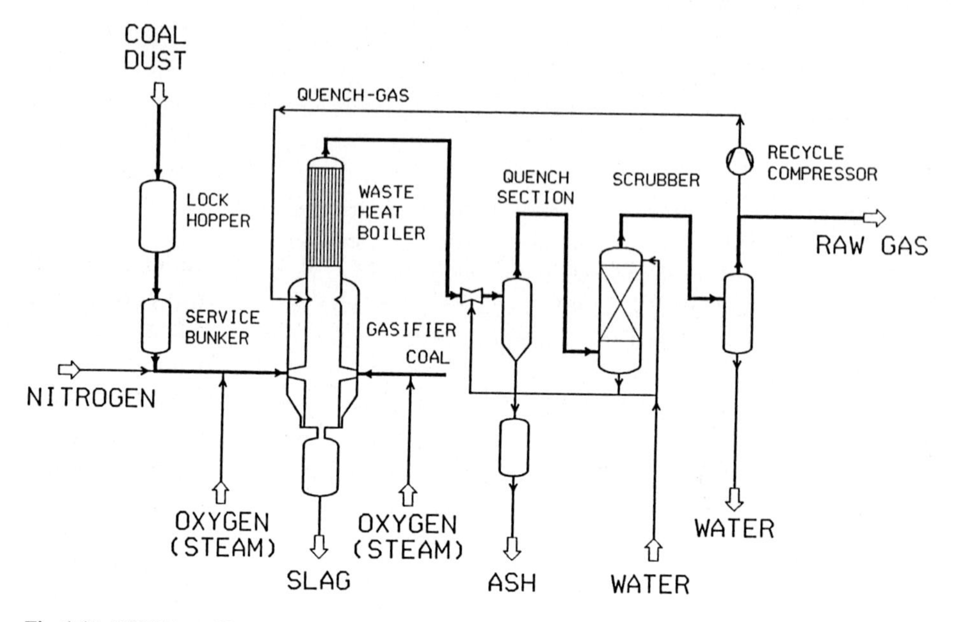

Fig. 1.15. SHELL gasification process (SCGP)

Table 1.10. Results of gasification of Illinois no. 6 bituminous coal by the *SHELL* coal gasification process

Coal Components (m.a.f.)	
C [% wt.]	77.3
H	5.9
S	4.3
N	1.4
O	11.1
Raw Gas Composition (dry)	
CO_2 [% vol.]	1.7
CO	61.6
H_2	30.6
CH_4	traces
N_2 + A	4.8
H_2 + COS	1.3
$\frac{m^3 (CO + H_2)}{1000 \text{ kg coal (m.a.f.)}}$	2090
$\frac{m^3 O_2}{m^3 (CO + H_2)}$	0.30
$\frac{\text{kg steam}}{m^3 (CO + H_2)}$	0.015

above the reaction zone, the hot reaction gas is quenched with recycled cold raw gas. The waste heat boiler serves to cool the raw gas and generate high-pressure steam. Fly ash is effectively intercepted in a venturi scrubber with downstream separator. Again, a sort of lock is used to discharge the ash. A scrubber followed by a separator to prevent carry-over of water mist into the raw gas line and hence into the recycle compressor cools the gas to its final temperature and removes the last ash particles.

Table 1.10 [1.1] shows the SCGP operating results for bituminous coal type Illinois No. 6 at a pressure of 24 bar. The cold gas efficiency under the conditions reflected in this table reaches 80.2 % for dried coal and decreases by about 0.8 % if the energy required for drying is considered. By-products from lignite gasification are essentially about 150 ppm of HCN and 20 ppm of NH_3; moreover, traces of almost all the elements contained in the coal are found also in the raw gas and the wash and waste water. The sulfur introduced with the coal appears in the raw gas at a rate of about 90 % in the form of H_2S and about 10 % as COS. Apart from some 800 ppm of methane, the raw gas does not contain any hydrocarbons. The output of the US plant, if bituminous coal type Illinois No. 6 is used, can meet the syngas demand of a 220 tpd methanol plant.

1.3.4 Slurry Gasification

As mentioned before, gasification of coal dust in a slurry with water is a special form of entrained flow[A gasification. As a matter of fact, the *TEXACO Coal Gasification Process* (TCGP) is the first pressurized version of the *Koppers-Totzek* technology that has been developed to the operative stage. Like SHELL, TEXACO benefited from experience with high-temperature gasification of gaseous and liquid hydrocarbons. The first pilot plant for a throughput of 15 tons per day was constructed as early as 1956. Work was discontinued, however, because of unsatisfactory operating results and because at that time the market was beginning to be glutted with petroleum. It was resumed some 15 years later, this time with better success [1.16]. Unlike SHELL, TEXACO retained most of the essential principles of gas/liquid gasification when they modified their system for coal gasification. The reactants are gasified in a downward cocurrent through the reaction space, and the proven feeding system for viscous liquids was retained at the expense of a significant burden on the cold gas efficiency – the coal is mixed with water and ground wet so that a coal slurry is obtained whose pumpability may have to be improved by suitable additives.

The proven quench system in the gasifier bottom, too, was retained with a few modifications to allow for coal as a feed and its ash content. Figure 1.16 shows the general configuration of a TCGP plant. Coal dust is introduced through a rotary feeder, mixed with water, and sent to the slurry service tank which is equipped with an agitator to prevent segregation. A reciprocating pump delivers the slurry at a constant rate by way of a burner tip into the reaction space. Oxygen is added through a concentric opening in the burner tip ensuring that it does not get into contact with the slurry until it has left the tip. Spontaneous evaporation

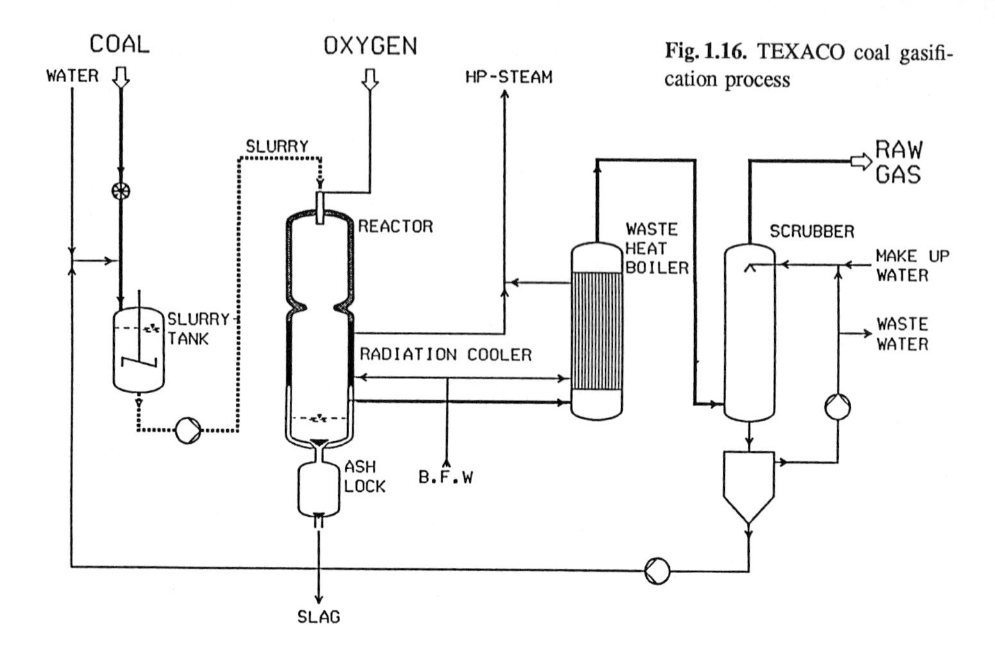

Fig. 1.16. TEXACO coal gasification process

of water from the slurry distributes the coal dust evenly into the oxygen flow to gasify it under entrained-flow conditions in the ceramic-lined reactor.

Evaporation and the resulting steam have a moderating effect upon the autothermal gasification reaction so that the maximum temperature in the reaction space hardly exceeds 1 500 °C. As shown by Table 1.11, the steam also shifts the water gas equilibrium significantly towards the CO_2 side in comparison with the SHELL gasification system. Together with the molten slag, the raw gas flows from the reaction space through a constriction into a radiant cooler in which the gas temperature is reduced to less than the ash sintering point and the partly solidified ash is eliminated. It is completely solidified in a water bath and discharged to the outside by way of a lock.

Thereafter, the gas flows through a waste heat boiler to generate high-pressure steam in the same way as in the radiant cooler. A scrubber operating partly on recycled and partly on fresh water cools the gas down to ambient temperature and removes the remaining ash particles. Freshwater make-up is necessary as some of the scrubber effluent has to be withdrawn to prevent a build-up of metal compounds from the coal in the scrubber loop. Most of the scrubber effluent, partly gas condensate and partly fresh water, is used to prepare the coal slurry.

Another version of the above process route may be conveniently used to produce ammonia or hydrogen. Rather than using the waste heat for high-pressure steam generation, the gas and the molten slag issuing from the reaction space are quenched to approximately 300°C. The steam so produced, together with the unreacted gasification steam, is sufficient to feed a CO shift conversion unit installed immediately downstream of the quench cooling stage, in which most of the carbon monoxide is shifted to CO_2, producing hydrogen along with it.

Table 1.11. Results of gasificaion of Illinois no. 6 bituminous coal by the *TEXACO* coal gasification process

Coal Components (m.a.f.)	
C [% wt.]	77.3
H	5.9
S	4.3
N	1.4
O	11.1
Raw Gas Composition (dry)	
CO_2 [% vol.]	12.26
CO	49.28
H_2	35.82
CH_4	0.36
N_2 + A	1.32
H_2S + COS	0.96
$\frac{m^3 (CO + H_2)}{1000 \text{ kg coal (m.a.f.)}}$	1970
$\frac{m^3 O_2}{m^3 (CO + H_2)}$	0.34
$\frac{kg\ H_2O \text{ (liqu.)}}{m^3 (CO + H_2)}$	0.21

If methanol syngas is to be produced in large plants, a combination of the two routes may be appropriate. As a rule, however, the first version, i.e. producing high-pressure steam rather than quenching the gas, if combined with a cold gas conversion unit (see Sect. 2.4) through which only some part of the raw gas may have to flow, will be the more economic solution.

The cold gas efficiency of this process, if Illinois bituminous coal is used, reaches almost 77 %. If lignite with its already high water content is used as a feed, the actual water content of the slurry will be much higher than the amount of water needed to prepare it and the cold gas efficiency will be reduced to clearly less than 70 %; hence, it is questionable on economic grounds whether the process can be employed for lignites having very high water contents. In 1987, four plants were in commercial service already, one of them in the USA with a coal throughput through one gasifier (1 stand-by) of approximately 34 t/h to supply syngas to a 500 tpd methanol plant and CO gas to a downstream acetic acid plant.

The latest in the row, which may in the meantime be considered operative as well, is the coal gasification process developed by *DOW Chemical Company*. A plant based on this process went onstream around the middle of 1987 in Louisiana. It provides for coal throughputs of about 83 t/h in the case of subbituminous coal feed, and of as much as 100 t/h if lignite is used. Although the gas

Table 1.12. Results of gasification of coal by the DOW coal gasification process

Type of Coal:	Subbituminous Coal	Lignite
Coal Components (m.a.f.)		
C [% wt.]	74.49	73.00
H	5.30	5.38
S	0.52	1.36
N	0.95	1.28
O	18.74	18.98
Raw Gas Composition:		
CO% vol.]	18.46	25.28
	38.46	33.74
H_2	41.35	38.82
CH_4	0.11	0.04
N_2 + A	1.48	1.82
H_2S + COS	0.14	0.22
m^3 $(CO + H_2)$ / 1000 kg coal (m.a.f.)	1900	1665
m^3 O_2 / m^3 $(CO + H_2)$	0.32	0.41
kg H_2O (liqu.) / m^3 $(CO + H_2)$	1.10	1.59

produced there is used as medium-BTU fuel gas, a glance at the figures in Table 1.12 shows that it could just as well serve as a raw gas for methanol production. The gas generator looks on the outside very much like a *Koppers-Totzek* or SHELL gasifier. Unlike these systems, however, the coal is fed to the reaction space not as a dry material but as a coal slurry. Some of the fuel is introduced together with the oxygen through diametrically arranged burners, and gasification in the reaction space takes place at temperatures of about 1 500°C. A portion of the coal slurry is injected into a second reaction zone disposed above the main one, to be gasified in an upward-flowing entrained bed. The quenching effect of this secondary slurry feed reduces the gas temperature sufficiently to prevent fly ash sintering on the tubes of the waste heat boiler provided downstream of the gasifier. The procedure of eliminating the fly ash and trim-cooling the gas is much the same as in the case of the SHELL process.

Completely new ground has been broken by DOW to discharge the slag which leaves the reaction space as a melt. After solidifying and cooling it in a water bath, it is ground – while still under pressure – and then let down to atmospheric pressure by a special system.

The gas output of the Louisiana plant (with one gasifier operating and one on stand-by) – if subbituminous coal is used as a feed – is sufficient to produce more than 1 600 tpd of methanol. The cold gas efficiency of the DOW process reaches 79 % from subbituminous coal and drops to about 72.5 % if lignite is used [1.17,18].

1.4 Interim Summary

The tables shown in this chapter might create the impression that some of the processes dealt with are unsuitable a priori for methanol syngas production on economic grounds because their $CO + H_2$ yields from the gasifier, in comparison with others, are significantly lower or even extremely low. Hence, it seems useful at this point, i.e. before going into gas cleaning and conditioning, methanol synthesis and distillation, and offsites, to sum up the results obtained so far, in order to arrive at a better assessment of the methanol efficiency of some processes. The comparison is based only on those gasification processes mentioned in chapter 1 for which all required figures were available from the same coal – Illinois bituminous No. 6 in this case [1.9].

Table 1.13 contains a summary of the following figures:

- The quantity of coal (m.a.f.) which has to be gasified in the gas generator to produce 1 kg of methanol. To derive these figures, the hydrocarbons and phenols leaving the gas cooling stage as liquids were considered in such a way that, with a reasonable loss in efficiency, they were mostly converted by partial oxidation into carbon monoxide and hydrogen. The ammonia condensed from the raw gas was left unconsidered. The methane produced by the LURGI dry-bottom gasifier and the LURGI/*British Gas* slagger was separated by pressure swing adsorption from the purge gas and then fed to an autothermal reformer to be reformed to $CO + H_2$. It was not overlooked at this stage that some 20 % of the purge gas from the synthesis loop have to be diverted to the fuel gas system to ensure that recycling via the autothermal reformer loop does not increase the nitrogen and argon contents in the recycled gas too much. These two processes are therefore at a slight disadvantage in this section because more *methanol potential* is exported from the process route in the form of CH_4 than in the case of the other processes for which a figure of 3 % of the $CO + H_2$ introduced to the synthesis loop was uniformly assumed to be lost by purge gas export.
- The quantity of coal which is needed to provide the energy required to produce and compress the oxygen needed for gasification.
- The quantity of coal needed to generate the process steam for the production of the $CO + H_2$ required in the methanol synthesis unit. This figure does not only allow for the steam that is fed to the gasifier and/or required for the partial oxidation of hydrocarbons and for purge gas reforming, but includes also the quantities needed for CO shift conversion to arrive at an H_2/CO ratio of 2 which is desirable for methanol synthesis, i.e. to produce hydrogen and reduce the CO content by the water gas reaction $CO + H_2O \rightleftarrows CO_2 + H_2$.

Table 1.13. Methanol efficiency of selected gasification processes

Type of Coal Gasification	LURGI Dry Bottom	LURGI/BG Slagger	KOPPERS-TOTZEK	SHELL	TEXACO	DOW
Coal to Gasifier [kg/kg methanol]	1.028	0.979	1.248	1.045	1.109	1.100
Oxygen [m^3/kg methanol]	0.526	0.469	0.88	0.655	0.743	0.72
Process Steam [kg/kg methanol]	1.380	0.604	0.496	0.477	0.430	0.230
Coal for Production of Oxygen and Process Steam [kg/kg methanol]	.241	.169	.258	.202	.219	.194
Pseudo - Methanol Efficiency [HHV Methanol/HHV Coal]	.542	.599	.406	.552	.518	.532

The last line of this table shows the totals of the above three figures. It is evident that the processes differ considerably as regards their methanol efficiency. However, this methanol efficiency must not be mistaken for a quasi process efficiency as a number of factors remained unconsidered which have to be taken into account for an assessment of the real efficiency. Such factors include

- steam generation from process waste heat and its direct use on a low level, e.g. for gas cleaning or methanol distillation,
- the energy and heat required to compress the raw gas and/or syngas and for all other power and heat consumers in the process plant and offsites.

A clearer impression can be gained from Chap. 7. A look at the process flow diagram of a complete coal-to-methanol plant (Fig. 7.1) and at the associated utility summary (Fig. 7.3) will show that 244 585 kg/h of coal, equivalent to 7 365 GJ/h of heat (HHV), are gasified in the gas generator section, while 1 363 GJ/h, equivalent to 44 832 kg/h of coal, are required to provide electricity (mainly for oxygen production and compression), steam and other utilities. The total coal quantity of 289 408 kg/h with a higher heating value of 30 114 kJ/kg is equivalent to 8 715 GJ/h of heat.

Chemical-grade methanol production amounts to 213 000 kg/h, equivalent to some 4 833 GJ/h on the basis of a higher heating value of 22 692.5 kJ/kg. Hence, a true process efficiency of 55.5 % can be calculated. If only the coal used for gasification were considered, the figure would be 65.6 %.

If these two figures are compared with the coal feeding rate for the LURGI dry-bottom gasifier in line one of Table 1.13 and with the total coal demand in the last line (*methanol efficiency*), it can be seen that the ratio between the total coal demand and the gasifier coal demand is approximately 1.25, i.e. almost the same as the above ratio of 1.18. One may therefore conclude that the simplified look at methanol efficiency is accurate enough to assess different types of coal.

2. How to Purify and Condition Methanol Synthesis Gas

2.1 Why and to What Extent Is Gas Purification Required?

All raw gases produced by gasifying coal require purification and, under certain circumstances, additional conditioning before they are used as synthesis gas for methanol synthesis. The task of the purification unit is to remove gas components which are detrimental to the process route or harmful to methanol synthesis, whereas gas conditioning has the purpose of providing the proper stoichiometric conditions for producing methanol from carbon monoxide, carbon dioxide and hydrogen, expressed by the formula

$$SN = H_2 - CO_2/CO + CO_2.$$

This condition, also termed the stoichiometric number, has its optimum for methanol synthesis at about 2.02 to 2.1, depending upon the concentration of inert gases such as methane, nitrogen, and argon.

The gas purification procedure downstream of coal gasification is by its very nature considerably more difficult and complex than purifying gases produced from gaseous or liquid hydrocarbons. Basically, it can be said that the fewer the molecules of the raw material from which synthesis gas is produced, the smaller will be the size of the gas purification and conditioning units. However, for the same raw material, namely coal, there are also considerable differences in the requirements for gas purification and conditioning, dependent upon the type of gasification processes applied.

In regard to purifying coal gases, the following impurities must be considered:

- Dust
- Carbonization products, tar, gas liquor
- Oxygen
- Cyanogen compounds
- CO_2
- H_2S and organic sulfur compounds
- Halogen and chlorine components
- Ammonia

In addition to the aforegoing components which are usually present in such quantities that they can be easily determined analytically, there are still a number of impurities which normally occur only in traces but can have a damaging influence on the copper catalyst in the methanol synthesis suction. They include all volatile metal compounds such as iron pentacarbonyl, as well as organic impurities with high molecular weights and double unsaturated hydrocarbons. The raw gases coming from the various coal gasification processes contain different forms and types of impurities and in addition often differ in pressure and temperature. These processes demand different types of gas purification and conditioning depending upon the type of impurity, the pressure and the temperature range as well as the steam content associated with them.

The requirements on purity and composition of the synthesis gas are the same for all modern low-pressure methanol processes in which copper-based catalysts are used. Synthesis gases are required to fulfill the following conditions:

- The stoichiometric ratio H_2–CO_2/CO + CO_2 must be greater than 2 and the gases are considered optimum if their stoichiometric ratios are as close to 2 as possible.
- Within the aforementioned stoichiometric ratio, the CO content should be as high as possible and the CO_2 content as low as possible but not less than 2.5 vol. %.
- The content of inert gases – all gas components except CO_2, CO and H_2 are considered as such – should be as low as possible.
- The content of sulfur compounds ought to be less than 0.1 ppm vol.
- Other impurities such as chlorine, HCN, NH_3 and double unsaturated hydrocarbons should not exceed the order of 0.1–3 ppm vol.

In addition, synthesis gases must be free of solvents used in the gas purification units themselves except if methanol is used as a solvent as it is the case in the Rectisol process, for example.

2.2 Principles of Gas Purification

The following types of processes may be taken into account for purifying and conditioning methanol synthesis gas derived from coal:

- Adsorption
- Absorption
- Mechanical/physical purification

2.2.1 Adsorption

The term *adsorption* describes the process by which impurities are collected on solid substances either due to physical depositing or by chemical reaction (chemosorption). The adsorption agents include activated aluminium oxide, iron hydroxide, zinc oxide, activated carbon and, to an ever-increasing degree today, molecular sieves which were usually developed for selective adsorption of individual gases or certain gaseous impurities. Wherever very intensive purification of the gases is required, i.e. where virtually all the impurities have to be removed, the adsorbent is not in a position to take up more than only fraction of its own weight of the substances to be adsorbed. Therefore, if large amounts of impurities have to be removed, it is necessary for economic reasons to use the adsorbent several times, i.e to regenerate it. Regeneration here means the expulsion of the components deposited on the adsorbent, and this usually takes place under heat and mostly with gases or steam.

Although gas purification systems of this type are generally applied only for removing small quantities of impurities and, above all, not for large gas throughputs, and therefore do not play an important role in purifying methanol synthesis gases from coal, the mechanism and basic operation of several adsorption processes will be briefly discussed here. Although most of the sulfur impurities are removed in various desulfurization stages, some of these stages do not reach the residual sulfur content of 0.1 ppm required for modern methanol processes. An adsorptive purification stage is therefore normally applied for post-purification purposes, using either adsorbents suitable for regeneration such as activated carbon or a nonregenerable adsorbent such as zinc oxide, dependent upon the quantity of sulfur remaining after the bulk removal stage.

2.2.1.1 Sulfur Removal over Activated Carbon

In our first example it is assumed that the sulfur compounds in the raw gas are removed from a large gas volume in a bulk removal stage to about 50 ppm only, and that an adsorption stage suitable for regeneration must therefore be employed for economic reasons. Adsorption processes in the molecular range are favored by increasing the partial pressure of the components to be adsorbed and reducing the temperatures. Two basic operations may therefore be used for adsorption, namely, a presure change or a temperature change, or a combination of these two.

In Fig. 2.1, the curve extending from Point 1 to Point 2, i.e. at constant temperature from P 1 to P 2, indicates the improvement in the loadability of the adsorbent as the temperature increases. The differential loading is G 2 minus G 1. The curve extending from Point 3 to Point 4, i.e. at constant pressure P 2 from T 3 to T 2, describes the effect that a temperature drop from T 3 to the lower temperature T 2 would have. Activated carbon was used as an adsorbent for this first example. Its adsorption potential is based on an extensive system of pores and can be adjusted for a variety of very different gas molecules simply by changing the pore size. Activated carbon processes normally operate with a combination of temperature and pressure changes.

This results in a curve extending from Point P 4 to Point 5. At pressure P 2 and a lower temperature T 2, the activated carbon is loaded, and at the lower pressure P 1 and higher temperature T 3 it is regenerated. Thus, regeneration proceeds above the operating temperature, i.e. heat has to be supplied. One recognizes that the loading differences from G 5 to G 4 is considerably higher than in the individual steps described previously. The progression of the curve

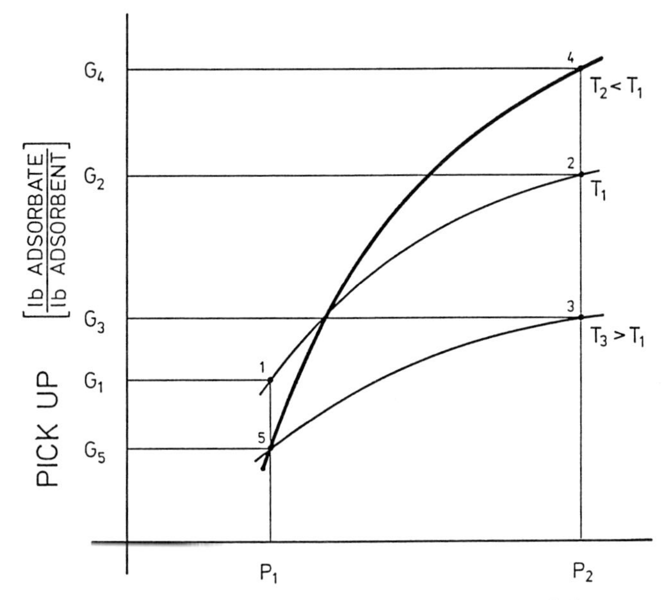

Fig. 2.1. Adsorption/desorption by temperature/pressure variation

shown is characteristic for each different adsorbent and cannot be calculated quantitatively in advance but must be determined empirically in all cases.

In practice, this sulfur removing stage then takes the form as shown in Fig. 2.2. The gas coming from the bulk sulfur removal stage flows through adsorber 1 where all sulfur components still contained in the gas are adsorbed to a residual quantity of 0.1 ppm. This adsorption may, for example, take place at pressure between 25 and 70 bar and at ambient temperature. The purified gas passes to the methanol synthesis. At the same time, the main gas route through adsorber 2 is closed, and this adsorber is depressurized to the regeneration pressure

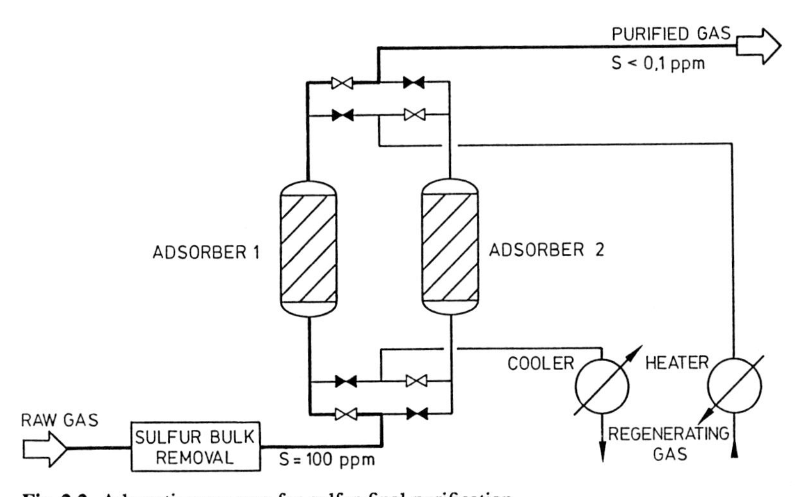

Fig. 2.2. Adsorption process for sulfur final purification

which ought to be below 2 bar. The depressurized gas volume leaving adsorber 2 and containg part of the sulfur adsorbed is lost for methanol production purposes unless at very high adsorbing pressures it is recycled to the sulfur bulk removal stage by means of a recompressor in order to minimize losses of valuable gas components.

Adsorber 2 is then purged with regeneration gas heated to about 260°C so that the impurities deposited on the adsorbent are removed almost completely. Finally, adsorber 2 is again brought to the operating temperature with cold regeneration gas and pressurized with purified gas to the process pressure. As soon as adsorber 1 has reached its load limit, a change-over is made to loading adsorber 2. The cycle time is specified according to the empiracally determined loadability of the adsorbent. Usually the loading time lasts about 12–24 hours and regeneration takes a very much shorter time. The change-overs from adsorption to regeneration and vice versa are generally automatic and time-controlled.

2.2.1.2 Sulfur Removal over Zinc Oxide

A second example will describe fine desulfurization over zinc oxide. Since zinc oxide can adsorb sulfur only by a chemical reaction forming zinc sulfide, regeneration will not be practicable. Although zinc oxide adsorbs some 20% of its own weight in the form of sulfur before the loading has an adverse effect on the residual sulfur content of the clean gas, its application is limited to gases containing only a few ppm of sulfur. This becomes clear by looking at the following figures:

If a gas flow of 100000 m^3/h, from which some 340000 tons of methanol could be produced annually, contained 20 ppm of H_2S, some 80 tons of zinc oxide would be required for H_2S removal.

Zinc oxide reaches its optimum loading at temperatures between 350 and 400°C. The raw gas is heated to this temperature range partly by heat exchange between the absorber inlet and outlet gases and partly by heat supplied for peak heating, and then fed to the absorber. In this way, H_2S is removed from the gas over zinc oxide to residual contents of less than 0.1 ppm. Organic compounds such as COS or mercaptanes occurring in the raw gas in concentrations of more than some 1–2 ppm have to be hydrogenated to H_2S before they can be adsorbed over zinc oxide. This hydrogenation normally takes place over nickel-molybdenum or cobalt-molybdenum catalysts preceding the zinc oxide absorber and operating in the same temperature range as the the H_2S adsorbers.

2.2.1.3 Pressure Swing Adsorption

Pressure swing adsorption (PSA), although included among the adsorptive gas purification processes and normally used to purify hydrogen, is used in the production of methanol from coal more as a gas separation process. The coal contains more carbon than needed for methanol production, and this carbon is removed from the gas flow in the form of carbon dioxide. Wherever possible and economical, it is therefore important that hydrogen should be recovered from offgases or residual gases. Gas conditioning practised in all methanol processes leads to a highly over-stoichiometric gas in the synthesis loop, i.e. a gas with a considerable surplus of hydrogen and a stoichiometric ratio of 5–7. Along with the inert gases extracted with the so-called purge gas, large quantities of valuable hydrogen are necessarily withdrawn from the synthesis loop.

Unless major quantities of methane in the purge gas speak in favor of reforming, the PSA process is normally employed to recover as much hydrogen as practical. This H_2 is then added to the synthesis gas which leaves the gas conditioning section with an appropriate surplus of carbon monoxide. The molecular sieve used for PSA is effective not only by its mini- and micro-pores but also produces electrostatic forces. This makes it particularly suitable for the adsorption of polar or polarizable gases while substances of low polarity such as hydrogen or helium flow through the molecular sieve practically without hindrance. In view of its short cycle times, the PSA system is especially appropriate for the removal of large quantities of impurities. If desired, it could, however, also be used to obtain exceptionally high hydrogen purities.

A complete PSA cycle consists of the following four basic steps:

- Adsorption
- Depressurization
- Purging at low pressure
- Pressurization

As described already for desulfurization over activated carbon, this cycle can be performed also in a two-adsorber system (although without the temperature swing practised there). Normally, however, systems with more than two adsorbers are used to reduce gas losses and increase hydrogen yields.

The combined action of several adsorbers as shown in Fig. 2.3 leads to the desired optimum. The four-adsorber system shown here also produces an even flow of depressurized gas which is frequently used for firing purposes and should

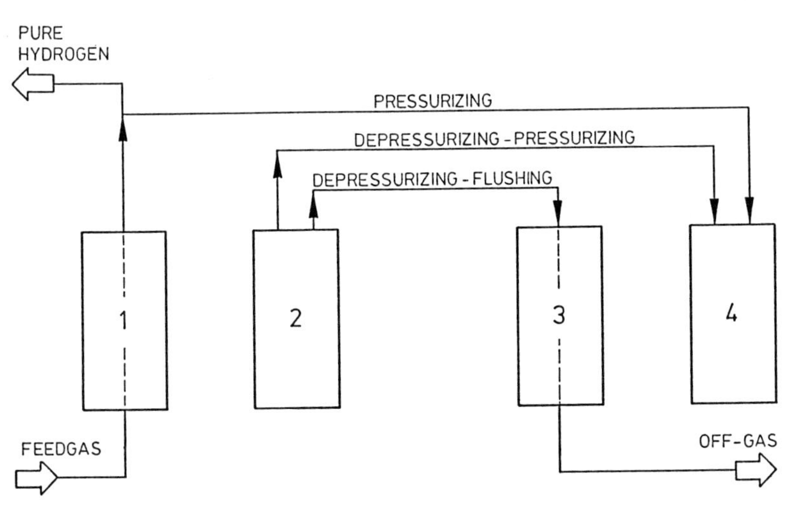

Fig. 2.3. Principle of pressure swing adsorption (PSA)

therefore not fluctuate too much in composition, pressure and flow rate. Adsorber 1 recovers hydrogen from the feed gas and feeds it to the loop at a uniform rate and at high pressure. Adsorber 2 is at the depressurization stage and supplies gas for the preliminary pressurization of Adsorber 4 and for purging adsorber 3 at low pressure. Hydrogen from the product piping is used to pressurize adsorber 2 to the feed gas pressure.

The hydrogen yield and capital investment are functions of the adsorption pressure and offgas pressure. As the adsorption pressure is normally determined by other process units – gas production and/or methanol synthesis – the adsorption system can usually be optimized only by way of the offgas pressure. Another factor influencing the hydrogen yield is the impurity content of the feed gas. The higher this content, the higher will be the hydrogen losses. Methanol production does not require highly pure hydrogen; rather, it is important to recover as much hydrogen as possible from a given feed gas – methanol synthesis purge gas in the present case. The influences produced by the adsorption pressure and the offgas pressure on hydrogen yields and their dependence on the hydrogen content of the feed gas are shown in Fig. 2.4.

Major advantages of the PSA process may be seen in the fact that hydrogen is obtained at a pressure which is only about 1 or 2 bar less than the feed gas pressure, that except for small quantities of instrument air, the PSA process does not need any utilities, and that the molcular sieves have an almost unlimited

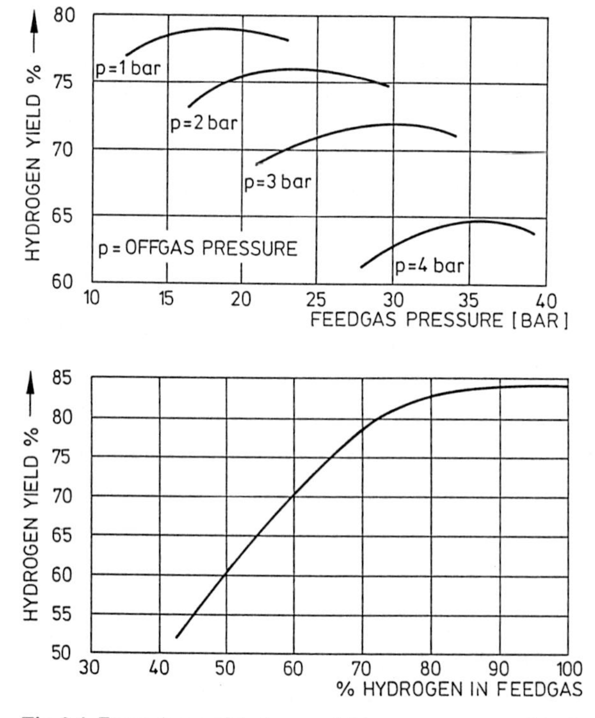

Fig. 2.4. Dependence of hydrogen yield upon pressure and hydrogen concentration in PSA

service life. Refilling the adsorbers is normally required only because frequent pressure changes produce minor abrasion losses at the molecular sieve.

2.2.2 Absorption

Whenever individual gas components are removed from gas mixes by means of a wash liquor (absorbent), this is termed *absorption*. Absorption processes range from physical processes to processes where absorption is of a purely chemical nature. Between these two extreme cases, there are a number of processes using a mixture of physical and chemical effects.

Physical wash processes are based on the so-called molecular attraction, while in *chemical wash processes* the substance to be removed reacts chemically with the wash liquor or some component of it.

Major applications of the absorption processes are the removal of acid gas components such as CO_2 and H_2S from gas mixes. Cases are not infrequent, however, in which other raw gas components such as higher hydrocarbons, trace impurities (metal carbonyls, NH_3) or steam are eliminated either alone or in combination with acid gas components. Absorption processes, which by their very essence are gas separation processes, are as a rule used for gas purification, i.e. undesired components are eliminated from a gas mix without significantly changing its pressure. Some absorption processes are, however, used to *recover* individual gas components. In these processes, the normally larger undesirable portion of the gas mix is left at absorption pressure while the desired gas component, after it has been desorbed, is usually obtained at ambient pressure.

Each absorption unit consists of two main parts, the *absorber* in which a solvent absorbs one or more gas components from a mix, and the *desorber* (regenerator) in which the absorbed gas is separated from the solvent again. Special cases in which a cheap solvent is loaded to exhaustion with substances which normally occur in the gas mix only in small quantities, and is then discharged, are becoming less and less applicable due to more and more stringent environmental protection laws.

With most absorptive gas purification processes, the eliminated raw gas components are desorbed again from the solvent in the same form in which they had been absorbed. Exceptions are the so-called *oxidation wash processes* in which the wash liquor reacts with the impurities so that the eliminated gas component is not obtained in the same form again. The *Giammarco-Vetrocoke Desulfurization Process* will be described here as an example to show the mechanisms by which absorption and desorption take place in these gas purification processes:

Absorption: $Na_3AsO_4 + H_2S \rightleftarrows Na_3AsO_3S + H_2O$

Desorption: $Na_3AsO_3S + 0.5\ O_2 \rightleftarrows Na_3AsO_4S$.

Exceptions from the rule that absorption and desorption are reversible include also hydrolysis processes, for instance, COS hydrolysis in amine solutions in which COS is reacted with H_2 to obtain H_2S and CO_2.

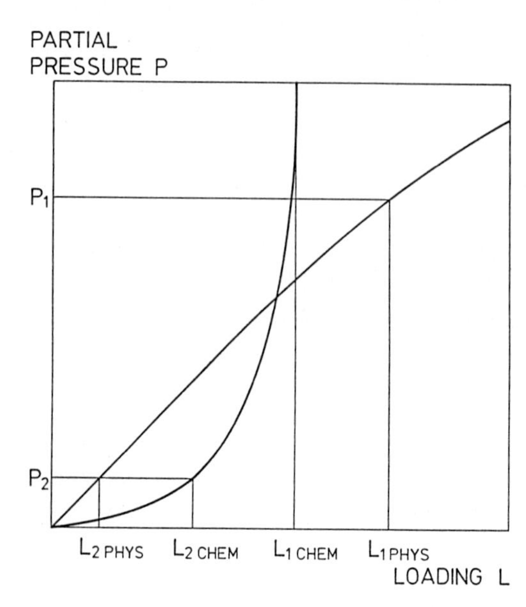

Fig. 2.5. Equilibrium of physical and chemical absorption

The *selection of the wash liquor* for a particular gas purification job is vitally dependent on the capital and operating costs – provided the wash liquor is suitable for the application at all. Both capital and operating costs are largely dependent on the quantity of circulating solvents and thus by the economics of the process. Fig. 2.5, showing the loadability of a physical wash liquor and of a chemical solution as a function of the partial pressure of the undesired gas component, suggests that chemical processes are more favorable at low pressures while physical absorption processes have an edge at high pressures. Both curves also show that a spent physical wash liquor releases a much greater part of the dissolved gas component simply by flashing (reducing the partial pressure) than a chemical wash liquor which can in almost all cases be completely regenerated only by an additional heat input (boiling). If in the case of physical absorbents the degree of regeneration achieved by flashing is insufficient, the residual load is stripped either by solvent vapours or by an extraneous stripping agent, or else by applying a moderate vacuum. Figure 2.6 shows these different regeneration methods.

Availability and utility costs are another important factor which influences the general decision in favor of a chemical or a physical gas purification process. Inexpensive waste heat at a suitable level favors chemical wash processes, low-priced electricity and cold cooling water have a favorable effect on the operating cost of physical wash processes. Other aspects which have an effect on the selection of the absorbent are its selectivity, the possible purification efficiency, and the properties and availability of the absorbent itself.

Selectivity is of interest under three aspects. For one thing, there is the capability of the absorbent to remove for instance one of two acid gas components down to a residual content of only tenths of ppm while the other is left in the

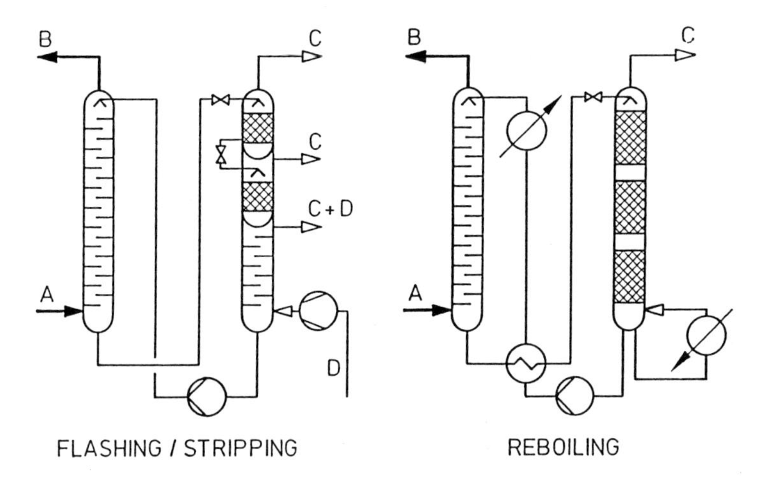

Fig. 2.6. Different types of regeneration of a solvent. (*a*) Feedgas; (*b*) purified gas; (*c*) offgas; (*d*) stripping agent

gas at a concentration of a few percent. This has particular significance for the production of methanol syngas; while H_2S – and of course all other sulfur compounds – have to be eliminated to less than 0.1 ppm because of the copper catalyst used in the synthesis loop, it is desirable to have a CO_2 content of some 2.5–3 vol. % in the syngas to ensure optimum catalyst activity.

Selectivity is also of major importance for the complexity of equipment required to treat waste gases wherever this may be necessary. A typical example illustrates the selectivity of a wash liquor regarding H_2S and CO_2. The higher the wash liquor can be loaded with H_2S while it absorbs only small quantities of CO_2, the higher will be the H_2S concentration in the acid gas (*Claus* gas) and the more favorable will be the conditions for sulfur recovery and tailgas purification. Thirdly, the selectivity of the wash liquor determines the quantities of valuable gases – carbon oxide and hydrogen in the case of methanol production – which leave the product gas flow together with undesirable components and are therefore lost for methanol production unless they are recovered by complex intermediate flashing of the offgases and recompressed into the raw gas. Regarding this third point, chemical wash liquors are clearly superior to physical ones. It must be admitted, however, that this disadvantage is in the case of many physical wash liquors offset to a considerable degree by the fact that their absorptivity for such gas components as H_2S, COS, CO_2, HCN and hydrocarbons, which are undesirable for methanol syngas production, increases steeply as solvent temperatures decrease, so that wash liquor demand becomes correspondingly small while the solubility for CO and H_2 remains nearly constant over a wide temperature range. Figure 2.7 illustrates this effect for methanol as a wash liquor.

Many chemical and physical wash processes – as well as combinations of the two – are suitable to remove undesired gas components to arbitrary residual contents. However, as the gas purity is largely determined by the number of

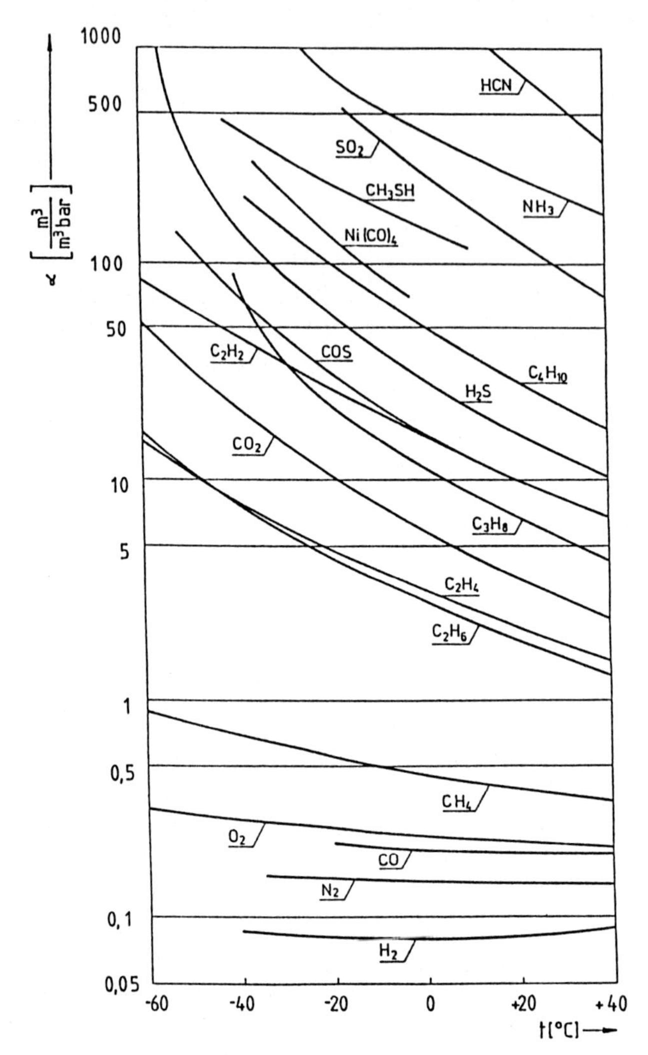

Fig. 2.7. Absorption coefficients of various gases in methanol

theoretical trays and thus by the overall height of the absorber – in addition to the wash liquor regeneration efficiency and the resulting wash liquor makeup rate – the gas purities required for methanol synthesis can be reached with economically justifiable methods only by very few processes.

Considerations regarding the wash liquor itself take account of its properties and availability. The corrosivity of wash liquors often increases dangerously as the pollutant loading increases and frequently calls for the use of special equipment materials or of corrosion inhibitors. Toxicity requires special measures in equipment design and process arrangements and not infrequently runs into considerable problems with environmental legislation. It must not be overlooked that toxicity does not in all cases describe an effect upon human beings, animals or plants, but may also include the poisoning effects of wash liquor traces on sensitive catalysts used in down-stream process stages. Such cases require secondary scrubbers or drying facilities in the clean gas flow. The term sensitivity as used here, describes the susceptibility of the wash liquor to damage from raw gas components or elevated temperatures. Such substances as oxygen, hydrogen cyanide, sulfur dioxide or even carbon dioxide as well as elevated temperatures during regeneration or the combined action of both effects may lead to degradation products which either have to be continuously eliminated from the liquor – in some cases with considerable effort – or, if this is not possible, require the entire wash liquor filling of a unit to be replaced very soon. The question of availability may be influenced by strategic considerations, embargos, or rapid procurement possibilities, for instance if major quantities of wash liquor are lost unexpectedly.

A number of appropriate summaries exist describing the suitability of wash processes and wash liquors in general [2.1] and for the purification of coal gas in particular [2.2]. The sections below will, however, describe only those processes which appear to be best suited to the purification of coal gas for methanol production.

2.2.3 Mechanical/Physical Gas Purification

Some processes to produce pure synthesis gas from coal require mechanical/physical gas purification. They involve, for instance, the collection of coal fines or ash particles, the interception of liquid byproducts (tar, oil, gasoline, water) by gravity separation. Such purification processes are designed exclusively to a specific coal gasification process and are not typical for the purification and conditioning of methanol syngas. They are therefore described together with the coal gasification process with which they are associated. Like the pressure swing adsorption described in Sect. 2.2.1, one modern mechanical/physical gas purification process, namely, membrane technology, has become important for the separation of purge gases from methanol synthesis.

2.2.3.1 Gas Separation by Membrane Systems

First attempts at separating gas mixes with moderate energy consumption date back to the beginning of this century. It was not until the time between 1960 and 1980, however, that membrane technology, which had originally been used exclusively to separate hydrogen from gas mixes, was considered also for other applications. This was also the time when plate-membrane systems were replaced by hollow-fibre membranes which are not only more compact and easier to manufacture, but offer the additional advantage that they can be used also under much severer operating conditions in large industrial plants.

The word membrane here stands for a thin polymer film which for the types used to today is characterized by a heterogeneous (asymmetric) structure, while the films used for filtration purposes have pores of even diameter all through the membrane. There is as yet no generally accepted scientific explanation of gas transport through polymer membranes. Attempts at describing these processes by equation systems lead normally back to *Fick*'s diffusion law and to *Henry's* law about the solubility of gases in polymer membranes. The design of membrane systems for specific separation applications is therefore based almost exclusively on empirical correlations between past operating data, and tests have to be made whenever new problems arise.

Although there are a number of materials with the desired pore structure, for instance silicone rubbers, hydrocarbon rubbers, polyesters, polycarbonates and others, their use for industrial applications is limited to polysulfones and cellulose acetates. While the latter have been used with good success for dehydration, technical gas separation relies exclusively on polysulfones which can be used up to approximately 70 °C (their melting point is around 200 °C) and at pressures between 15 and 140 bar. The lowest pressure differential between the feed gas side and the permeate gas side is 3 : 1 and this differential pressure determines the wall thickness of the membranes. Figure 2.8 shows the design of a membrane element developed by *Monsanto* Company, USA and marketed by the name of *Prism® separator*. Each of these elements or modules contains thousands of hollow fibres packed to a density of approximately 100 per cm^2.

Various gases differ by their rates of permeation, depending on their capacity of dissolving in the hollow fibre membranes or of diffusing through them. The *rapid* gases include H_2O, H_2, H_2S and CO_2 in this order, and the series contin-

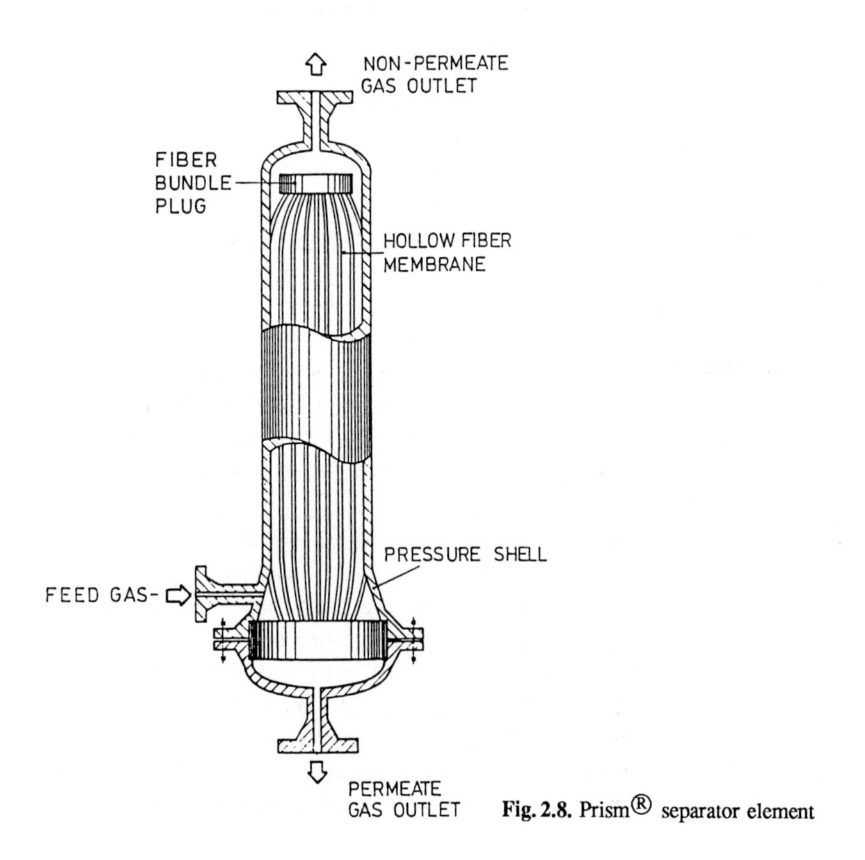

Fig. 2.8. Prism® separator element

ues towards the *slow* side by O_2, Ar, CO, CH_4. An example may illustrate how a membrane system separates a purge gas from methanol synthesis into undesired components and components whose recirculation is worthwhile. Purge gas from the methanol synthesis unit operating on synthesis gas from high-temperature coal gasification which contains very little methane is available at 67 bar and 35 °C with the following composition:

CO_2	3.78 vol. %
CO	4.34 vol. %
H_2	55.76 vol. %
CH_4	0.26 vol. %
N_2	29.11 vol. %
Ar	6.21 vol. %
H_2O	0.01 vol. %
CH_3OH	0.53 vol. %

A simple water scrubber is used as a preliminary stage to the Prism separators to remove all but traces of the methanol from the purge gas which might otherwise damage the membrane material. In order to avoid residual water condensation in the membrane system, the feed gas is then heated to some 10 °C above the water dew point and then fed to the Prism separators. The permeate gas, i.e. the gas that has permeated the membrane, consists of a mixture containing some 90 % of the hydrogen in the feed gas and has the following composition:

CO_2	4.64 vol. %
CO	2.56 vol. %
H_2	78.50 vol. %
CH_4	0.09 vol. %
N_2	11.28 vol. %
Ar	2.81 vol. %
H_2O	0.12 vol. %

This gas is obtained at a pressure of almost 22 bar so that little energy is needed to recompress it into the syngas. The pressure of the other part of the feed gas drops by not more than 0.5 bar and the gas is thus obtained at a pressure of 66.5 bar.

The hydrogen purity can be considerably increased if the user is satisfied with a hydrogen yield lower than that in the above example. Reducing the pressure on the permeate gas side may lead to higher hydrogen yields or hydrogen purities. It is well conceivable that this process, which was introduced by the chemical industry only a few years ago, still offers a considerable development potential and that it may be worth considering its use also at other points of the *coal-to-methanol* process chain.

2.3 Different Types of Gas Purification Processes

2.3.1 Physical Purification Processes

2.3.1.1 The Rectisol® Process

Developed around the middle of the fifties by LURGI and matured to industrial application together with *Linde*, the *Rectisol Process*, which had originally been tailored to the specific requirements of purifying coal gas, is still one of the most effective and flexible gas purification processes today [2.3,4]. Methanol, which is used as a wash liquor, is distinguished not only by its high selectivity between sulfur components and CO_2, but also by a high absorptivity at low temperatures and convenient regenerability at temperature levels where cheap waste heat can be used. Unlike other physical wash liquors such as n-methylpyrrolidon (NMP) or mixes of glycolethers whose application is limited to temperatures above 0° because of their high viscosity, methanol is characterised by very low viscosity even at low temperatures so that full advantage can be taken of its absorptivity, which increases with decreasing temperatures down to $-75\,°C$. Methanol seems to be particularly suitable to purify methanol syngas because

- the solvent is the same as the product of the plant,
- methanol enables not only sulfur as the major impurity including its high molecular compounds to be removed to residual contents of less than 0.1 ppm but – as shown by a look at Fig. 2.7 – also has a very high absorptivity for many other catalyst poisons such as metal carbonyles, acetylene and hydrogen cyanide,
- residual wash liquor need not be removed from the purified syngas – for instance by a downstream scrubber – since methanol entering the synthesis loop cannot produce any malfunction or damage.

A Rectisol unit may either be of *standard type* (all-out type) removing all undesired components together, or of the selective type in which essentially all sulfur components and metal carbonyles, hydrogen cyanide and unsaturated hydrocarbons but as little CO_2 as possible are removed, while most of the carbon dioxide is washed out in a second stage which is in some cases arranged downstream of the CO shift conversion unit. Which of the two systems will be more appropriate in a particular case depends on the type and quantity of the undesirable components in the coal gasification gas, on the requirements regarding acid gas composition, and on the required purity of the offgases to be discharged to the atmosphere.

The equipment of an all-out purification unit is certainly less complex than that of the selective type. However, since the combined removal of the sulfur components and the entire CO_2 would leave the acid gas with so low a sulfur load that sulfur recovery in a *Claus* unit would not be possible any more, this arrangement requires a considerable effort to desulfurize the offgases sufficiently to conform to environ mental legislation and to make sulfur recovery cost-effective.

Rectisol gas purification for sulphur and CO_2-removal

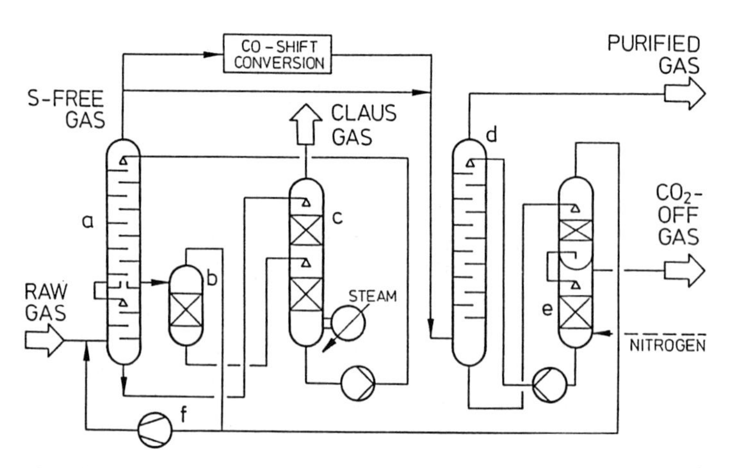

Fig. 2.10. Selective rectisol gas purification system. (*a*) H_2S absorber; (*b*) flash stage; (*c*) hot regenerator; (*d*) CO_2 absorber; (*e*) flash/stripp regenerator

from medium temperature gasification are typically distributed to the gas flows leaving a purification unit of a type which can be constructed with reasonable expenditure.

A selective Rectisol unit as shown in Fig. 2.10 in conjunction with slip stream CO shift conversion is not only more flexible but more cost-effective as well. It will be preferable above all to purify gases from high-temperature coal gasification whose low CO_2 contents in the raw gas normally do not require the sulfur content in the *Claus* gas to be increased by reabsorbing H_2S.

Figure 2.10 shows that the raw gases are first fed to the H_2S absorber where all sulfur components, metal carbonyles, hydrogen cyanide and double unsaturated hydrocarbons are removed with heat-regenerated methanol at temperatures around $-30\,°C$. Some impurities such as carbonyles and HCN are much better soluble in methanol than sulfur compounds so that the regeneration effort and /or stripping steam demand in the hot regenerator is correspondingly higher for these substances. This is why such substances are normally eliminated in a small prepurification stage which needs only a fraction of the methanol required for sulfur absorption. The impurity-laden methanol from this prepurification stage is then flashed into the top of the hot regenerator where the stripping steam/wash liquor ratio is several times greater than in the bottom section of the vessel. The laden methanol from the main purification stage is first let down to about half the operating pressure of the absorber, and the valuable CO and H_2 gases so released are recompressed into the raw gas. Thereafter, the methanol is fed to the upper half of the hot regenerator to be completely regenerated by boiling it out with steam. The entire heat- regenerated methanol is cooled down in heat exchangers and pumped to the top of the H_2S absorber from which a gas with a residual sulfur content of less than 0.1 ppm is delivered to the CO shift conversion unit.

Only a slip stream of the sulfur-free gas is normally put through the shift conversion unit because it is more cost-effective to convert a slip stream of a gas containing for instance 60 % CO to 5 % residual CO than converting the entire gas throughput to 28–30 % as required for the desired stoichiometric ratio.

A sulfur-free, partly converted gas is then fed at approximately −60 °C to the CO_2 absorber where enough CO_2 is removed to bring the carbon dioxide content of the gas down to the level required for a stoichiometric ratio H_2–CO_2/CO + CO_2 of approximately 2.05 as is it desirable for methanol synthesis. Since the residual CO_2 content in the methanol syngas – as described in Sect. 2.1 – should not be less than 2.5 vol.%, the CO_2 absorber usually has a much smaller number of trays than the H_2S absorber. While the latter is equipped with some 80 trays, less than half that number are required in the CO_2 absorber. The high residual CO_2 content also facilitates regeneration of the impurity-laden methanol from the CO_2 absorber. For gases from medium-temperature coal gasification, whose CO_2 content does not reach the optimum of about 28–30 vol.% and whose CO content may have to be increased to as much as 6 % in order to reach a stoichiometric ratio of 2.05, flash regeneration of the laden methanol will in most cases be sufficient. At least two stages are provided for this purpose, the pressure in the first flash stage being adjusted approximately to the CO_2 partial pressure above the laden solvent. The high-H_2 and high-CO flash gas is then recompressed together with the flash gas from the H_2S scrubber section. The second flash stage is normally operated at atmospheric pressure. As the raw gas is desulfurized to very low residual levels, the CO_2 offgas from this stage contains less than 1 ppm total sulfur, which is no problem for the environment.

If the CO_2 content in the clean gas has to be adjusted to considerably less than 6 %, this can be done only by either flashing the solvent into a slight vacuum in another regeneration stage following the atmospheric stage or by stripping it with nitrogen. Nitrogen stripping is normally more economical in coal-based methanol plants as such plants have in most cases air separation units associated with them which not only provide the oxygen required for gasification but can also deliver large quantities of nitrogen. Whereas the chilling required for a Rectisol unit – resulting from insulation losses, cold losses with clean and offgases and pumping heat transferred to the solvent – is provided at as high a level as possible, usually around −30 °C, at a suitable point in the methanol loop, CO_2 desorption in the flash stages, with a sensible heat of some 17 000 GJ/kmol, cools the methanol down to temperatures around −60 °C which are desirable for absorption.

Some types of coal contain so little sulfur that in spite of low CO_2 contents the acid gas from the hot regenerator is not suitable for the *Claus* process.

Measures in such a case may either be the same as described above for an all-out purification unit, i.e. increasing the sulfur content, secondary acid gas scrubbers, etc., or the *Claus* unit may be operated with oxygen or oxygen-enriched air. This will be appropriate for sour gases containing between 8 and 25 vol. % sulfur.

In order to keep methanol losses down, the atmospheric offgases will either have to be treated with water over a few wash trays, or their methanol will have to be condensed by cooling them to very low temperatures. Normally, both the raw gas and the converted gas contain steam which is absorbed by the methanol in the wash unit. To prevent water build-up in the loop, which would impair the absorption capacity of the methanol, a small methanol quantity is continuously withdrawn from the loop and fed to a water/methanol distillation column together with the water/methanol mix from the after-scrubbers. In this column, the methanol is boiled off and recycled to the wash loop. The water drained from the column bottom also contains all the components with boiling points higher than methanol which were dissolved in the absorption stages, as well as any solid impurities entrained with the raw gas. As a general summary it may be said that in a two-stage purification unit, methanol as a solvent with a boiling point lower than that for water has the advantage that all impurities whose boiling points are lower than that of methanol will get into the *Claus* gas while the high boilers and solids are collected in the waste water.

For clearness sake, Figs. 2.9 and 2.10 do not show the after-scrubbers, the water/ethanol distillation equipment and the chilling facilities.

2.3.1.2 The Purisol® Process

NMP or n-methyl-2-pyrrolidon, as it is called more completely, is one of the most universally applicable solvents [2.5]. Its gas purification potential was recognized by LURGI, the process licensor, as early as 1960, and this was also the time when the first patent for the Purisol Process was granted [2.6,7]. Two of the most attractive features of NMP are its high H_2S absorptivity at ambient temperature and its high selectivity for H_2S over CO_2. Since NMP dissolves almost ten times as much H_2S as CO_2, Purisol gas purification units can produce a high-H_2S *Claus* gas even from low sulfur coals. The solubility of hydrogen and carbon monoxide in NMP is similar to that in methanol so that Purisol units cannot do without recompression of the regeneration gases either. NMP vapour pressure is low; only small quantities of vapourous NMP are therefore lost from the absorber together with the clean gas at the pressure levels of some 30–100 bar at which the Purisol process normally operates. As in the case of all other wash liquors described below, an after-scrubber is nevertheless required in methanol plants to prevent damage to the methanol catalyst.

Figure 2.11 shows *Henry's* absorption coefficients for CO_2, H_2S and COS at atmospheric pressures for a number of solvents plotted as a function of temperature. The figure shows clearly that the H_2S absorptivity of all other wash liquors is greater than that of methanol while their COS absorptivity is somewhat poorer. As regards CO_2, all solvents behave more or less alike. The phyiscal data listed in Table 2.2 for a number of physical wash liquors show that, like glycolether, NMP is considerably more viscous than methanol. All data in this table refer to water-free solvents. Actually, extrapolation of these figures suggests that glycolether has the same viscosity already at about 40 °C, and NMP at 0 °C, which methanol reaches only at a temperature of −50 °C. As mass transfer becomes more difficult and the pressure losses within the solvent loop increase as the viscosity increases, the applicability of the Purisol process, like the other

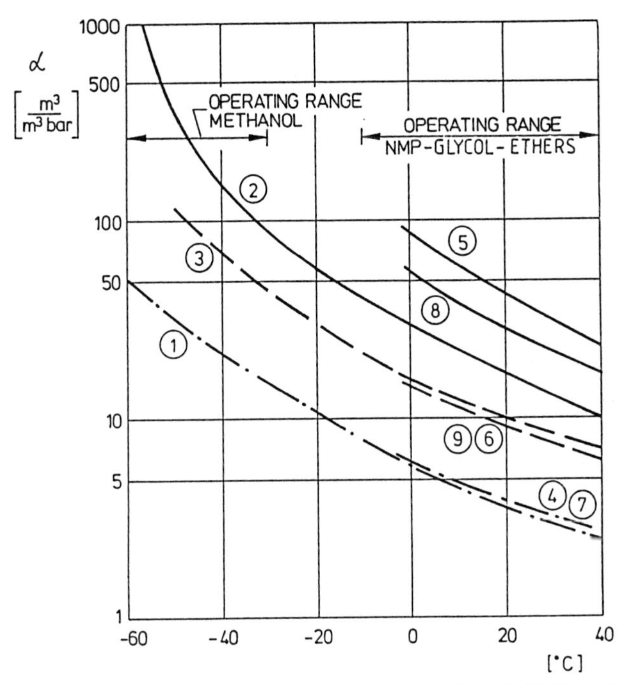

Fig. 2.11. HENRY's absorption coefficients for 1. CO_2 2. H_2S 3. COS in methanol 4. CO_2 5. H_2S 6. COS in NMP 7. CO_2 8. H_2S 9. COS in glycol ethers

Table 2.2. Properties of physical solvents

		Methanol	n-methyl-2 pyrrolidone	Dimethyl polyglycol-ether
Formula $CH_3O(C_2H_4O)_xCH_3$		CH_3OH	C_5H_9NO	
Mol. weight		32	99	178 - 442
Boiling point, 760 Torr	°C	64	202	213 - 467
Melting point	°C	-94	-24.4	-20 ÷ -29
Viscosity	cP	0.85/-15°C	1.65/30°C	4.7/30°C
		1.4 /-30°C	1.75/25°C	5.8/25°C
		2.4 /-50°C	2.0 /15°C	8.3/15°C
Spec. weight	kg/m^3	790	1.027	1.031
Heat of evaporation	kJ/kg	1090	533	
Spec. heat, 25°C	kJ/kg°C	0.6	0.52	0.49

processes using glycolether which have been described below, is limited by economic factors to a range between −10 and +40 °C. Exceeding these limits on the low temperature side would lead to a very large number of mass transfer trays and thus to a technically impossible absorber height as well as to a higher energy consumption for the wash liquor pump, while on the high temperature side the sharply reduced absorptivity of the solvent would require the wash liquor circulation to be increased considerably.

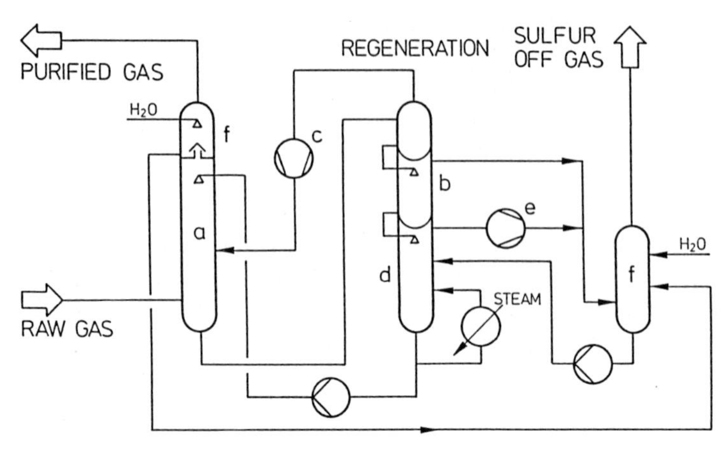

Fig. 2.12. Purisol gas purification process. (*a*) Absorber; (*b*) flash stages; (*c*) recompressor; (*d*) hot regenerator; (*e*) vacuum blower; (*f*) solvent recovery

The low vapour pressure of NMP – Table 2.2 shows that it has a boiling point of 202 °C at atmospheric pressure – would require heat regeneration above this temperature level. However, since NMP begins to decompose at temperatures above 180 °C, it has to be regenerated either at vacuum or with the aid of a stripping agent. Water can be used only very conservatively in this case as it impairs the absorptivity of NMP. CO_2 is often available at slightly elevated pressure from the regeneration section of the CO_2 purification unit, but stripping is practicable only if the raw gas contains only very little sulfur and the sulfur offgas can be incinerated.

Figure 2.12 shows the arrangement of a selective H_2S Purisol unit used to eliminate sulfur from a coal gas containing not only hydrogen, CO and a minor quantity of methane, but also 4 % CO_2, 1 % H_2S and 400 ppm of other sulfur compounds, mainly COS. The raw gas is purified in the absorber to residual contents of 1 ppm H_2S and approximately 200 ppm COS. As shown by Fig. 2.11, COS solubility is some fives times lower than H_2S solubility. Designing the unit for residual COS contents around 1 ppm would therefore require five times as much solvent and the process would no longer be cost-effective.

Solvent residues are removed from the purified raw gas in a secondary scrubber. The spent solvent is flashed to atsmopheric pressure in two stages and the gases released in the first stage are compressed and recycled to the absorber. The NMP, which thereafter contains only minor quantities of impurities, is then boiled in a third regeneration stage in a vacuum at 140 °C and regenerated completely. A vacuum blower compresses the gases from the vacuum stage to a pressure slightly above atmospheric before they are added to the flash gas from the secondary regeneration stage and fed to a water scrubber which mainly serves to recover the solvent. This arrangement also ensures that no additional NMP drying facilities are required as the water balance of the scrubber can be adjusted in such a way that all the feed water, except that part of it which is entrained with the clean gas will be discharged together with the sulfur offgas. This keeps the water content of the solvent down – normally to less than 1 wt.% – so that the NMP retains its full absorption capacity and corrosion resulting from the interaction of acid gases and high water contents is avoided.

Owing to the above-mentioned high selectivity of NMP, the sulfur off-gas contains about 50 % sulfur components and 50 % CO_2 and can therefore be treated by any standard *Claus* unit. The clean gas, as already mentioned, contains some 1 ppm H_2S and 200 ppm COS. This purity is insufficient for methanol synthesis so that additional fine desulfurization is required. This can be conveniently achieved by providing a COS hydrolysis stage upstream of the Purisol unit and an adsorption stage downstream of it. In what is termed the COS conversion unit, some 80 % of the COS are hydrolized to H_2S (see Sect. 2.3.4) so that only 80 ppm of COS enter the Purisol unit and 40 ppm leave it again. This is an order which makes adsorptive removal possible whereas a sulfur content of 200 ppm which would be normal without COS hydrolysis would lead to unjustifiable adsorbent consumption rates. Since COS is not very readily adsorbed over zinc oxide, the adsorption unit has to consist of two stages – a hydrogenation stage in which the COS is hydrogenated to H_2S over a nickel or cobalt molybdenum catalyst, and an adsorption stage in which the H_2S is adsorbed by the zinc oxide to less than 0.1 ppm.

Gases produced in high-temperature coal gasification units require in almost all cases some of the CO to be converted and the resulting CO_2 to be removed in order to adjust them to the appropriate stoichiometric ratio. Normally, only a slip-stream of the gas is needed for CO shift conversion (see Sect. 2.4). If the Purisol process is used for downstream CO_2 removal, it may be worth considering adsorptive fine desulfurization without COS conversion upstream of the H_2S scrubber. With this configuration, about half the gas rate is put through the CO shift conversion stage in which CO shift conversion is accompanied by an almost quantitative conversion of COS to H_2S. The converted gas then enters the CO_2 purification stage with the residual COS content of approximately 1 ppm. In this stage, the H_2S is eliminated to a residual content of 1 ppm while the COS passes the unit almost unaffected. When the gas flow from the CO_2 removal stage is combined with the gas that has bypassed the CO shift conversion stage, the mix contains only about half as much sulfur as at the outlet from the H_2S removal stage, and zinc oxide consumption may be tolerable. It must not be overlooked, however, that the CO_2 offgas from the CO_2 removal unit then contains more than 600 ppm of H_2S and the offgas has to be fed to an incinerator to burn H_2S to SO_2 before it can be discharged to the atmosphere.

Another possibility of keeping the expenditure on fine desulfurization units within reasonable limits is the addition of 1,2-dimethylimidazole [2.8] to the NMP. This additive has the effect that almost all the COS entering the H_2S elimination stage is hydrolized to H_2S and CO_2. Instead of the 200 ppm mentioned in the first example, only 5–10 ppm of COS would in this case leave the H_2S wash unit while the balance would be removed in the form of H_2S.

Elimination of CO_2 by means of NMP may be attractive under certain conditions, for instance where inexpensive electrical energy is available and cheap waste heat, which would favor the use of a chemical wash process, is unobtainable. A certain importance is often attributed also to the use of the same solvent for the removal of sulfur and CO_2 as it offers the advantages with respect to storage and recovery. The design of a CO_2 wash unit using NMP as a solvent is largely identical with the CO_2 removal section of a Rectisol unit as shown in Fig. 2.9. Owing to the fact that the CO_2 absorptivity of NMP is poorer than that of methanol, its regeneration is easier and can in many cases do without stripping gas simply by flashing the solvent to reach residual CO_2 contents of around 5 % in the clean gas.

In Purisol units, the high CO_2 desorption heat is used also to reduce solvent demand. The NMP entering the absorber top at a temperature of approximately 3 °C and heating up on its way to some 45 °C is cooled to approximately 3 °C by a loop installed in the absorber bottom. This means that it picks up more CO_2 in the bottom section of the column and a considerable amount of heat is removed from the unit without needing a refrigeration system. When it is flashed, the solvent then cools down from 30 °C to about 0 °C. The energy fed to the pump delivering the NMP to the absorber heats up the solvent to 3 °C – the temperature at which it enters the absorber.

The Purisol process can be used also for what has been termed the all-out purification process required to produce methanol from gases resulting from medium temperature gasification of coal, which contain relatively little CO but more CO_2 than required. However, if NMP is to be used for such applications, it will be necessary not only to increase the sulfur content of the sulfur offgas, but also to provide for additional fine desulfurization.

In the case of large plants it will often be worthwhile to install a refrigeration unit, which enables the NMP to be used at temperatures below 0 °C so that the solvent circulation can be reduced. For further energy saving, the high pressure solvent pump is sometimes combined with an electric motor and an expansion turbine so that about 50 % of the power demand for the pump are provided by the expansion turbine.

2.3.1.3 The Selexol Process

The *Selexol Process*, which had orginially been developed by *Allied Chemical Corporation* to remove the CO_2 from NH_3 synthesis gas, uses a mixture of high-boiling polyethylene glycol/dimethyl-ethers as a solvent. Although it differs considerably in some of the physical aspects, Selexol exhibits similar properties for gas purification as the NMP described above. Both solvents are therefore used also in similar applications. The use of glycolether at temperatures around 0 °C suffers from its high viscosity, which is about four times as high as that of NMP, and from its high boiling range (see Table 2.2). Conditions for the use of the Selexol solvent can be improved by adding water, which reduces both its viscosity and boiling range.

The addition of water has a less adverse effect on the absorption capacity than in the case of NMP. The reason lies in the high molecular weight of the glycolethers. If for instance 1 wt.% of water is added to the Selexol solvent, this has no significant effect upon its absorbtivity, but it reduces the boiling range considerably because 1 wt. % of water – assuming an average molar weight of the glycolethers of 310 – already accounts for 17 % of the molar weight. In the case of NMP, on the other hand, 1 wt.% of water is equivalent to only about 5.5 mol.%. If the H_2S content in the syngas has to be reduced to less than 1 ppm – as required for methanol production – the Selexol process, too, cannot do without heat regeneration under vacuum.

The Selexol solvent can be used under similar conditions as NMP also to remove CO_2. Formerly, the CO_2-laden solvent was regenerated by flashing it in several stages and then stripping it with air. This is no longer practised today, because the presence of oxygen led to the precipitation of elemental sulfur in the vessel by way of the so-called *cold Claus reaction* from H_2S-laden solvents, even though H_2S was present only in the order of a few ppm. If residual CO_2 contents of less than 3.5 vol.% are required, the solvent is finally regenerated under vacuum. The poor COS solubility of the glycolethers makes an additional fine desulfurization stage indispensable for the Selexol Process [2.9].

The high vapour pressure of the glycolethers has the effect that only very small solvent quantities leave the system together with the clean gas and with the sulfur and CO_2 offgases, whose recovery in a secondary scrubber does not seem economically justifiable at first glance. However, even traces of this solvent have to be eliminated from the clean gas before it can be used for methanol synthesis so that an after-treatment is necessary at least at this point.

If no secondary scrubbers are provided in the offgas streams, the Selexol solvent losses with the offgases, however small they may be, have the effect that first of all the lower boiling glycolethers are lost as they are topped off so to speak. This means that the average molecular weight of the solvent rises consistently and the viscosity and the boiling point increase steadily. Under certain circumstances, the entire solvent filling of the plant may have to be replaced. The high boiling range of the glycolethers makes steam heating of the solvent dryers impossible. Fuelgas heated glycol reconcentration units are therefore used to remove water from the solvent [2.10]. Refrigeration units and expansion turbines are frequently considered also for Selexol units, the former to reduce wash liquor circulation and equipment size, the latter to reduce the power demand for the solvent circulation pump.

2.3.1.4 The Sepasolv Process

In 1976, BASF offered an absorptive gas purification process which is marketed today under the name *Sepasolv MPE* [2.11] and which uses a solvent similar to that of the Selexol process, i.e. methyl-isopropyl-ethers of polyethylene glycols. The applications of this process may be assumed to be similar to those of Selexol. Preference is given to ethers with 3 to 5 ethylene glycol groups and an average molar weight of 315 kg/kmol. The percentage of low-boiling components is small so that there is hardly any risk for the solvent composition to change noticeably because low boilers are expelled.

The H_2S and CO_2 solubilities are similar to those in Selexol (Fig. 2.11) and so are the achievable residual contents of both components. Up to 3 wt.% of water are added to the Sepasolv solution to adjust its viscosity and boiling range. This reduces the H_2S solubility to about two-thirds of that in water-free Sepasolv, while CO_2 and COS solubilities are only slightly affected. Sepasolv has a vapour pressure at 30 °C of only about 10^{-6} bar; no water scrubbers are therefore required to recover solvent vapours from the offgases. The licensor claims that Sepasolv, unlike Selexol, is in no way corrosive. With respect to the desulfurization of natural gas, this seems to be true, while for synthesis gases produced from coal, where the corrosion hazard in the CO_2 purification stage is much higher because of the high CO_2 partial pressure, this has yet to be confirmed; no such plant has so far been operating on an industrial scale.

2.3.1.5 Other Physical Purification Processes

In addition to the processes described above, there are a number of other physical absorption processes which will not be discussed here in detail either because they can be reasonably used for the purification of methanol syngas from coal with economically justifiable expenditure only in exceptional cases, or because they have not yet been tested on an industrial scale. These processes include

- the *Pressurized Water Process* – probably the oldest of all known purification processes. Because of its poor selectivity and high energy consumption it can be used, if at all, only to remove CO_2, provided that very cold water is available in sufficient quantities
- the *Fluor Solvent Process*, which uses propylene carbonate as a solvent [2.12]. The absorptivity of this solvent for CO_2 and sulfur components is much poorer than that of the other solvents described in detail above; above all, the selectivity between H_2S and CO_2 decreases consistently with rising gas pressures. The *Fluor* solvent process has so far been used only to purify natural gases
- the *Estasolvan Process* developed by *IFP/Uhde-Hoechst* which uses tri-n-butylphosphate as a solvent [2.13]. The solubility of H_2S in TBP is similar to that in propylene carbonate. The H_2S selectivity of TBP is very high, although at the expense of a low absorptivity for CO_2. Nothing has been published so far about the operating experience with this process for the purification of synthesis gas.
- the *Leuna Gaselan Process* which was developed at the *VEB Leunawerke Walter Ulbricht* and which uses n-methyl caprolactam (NMC) as a solvent [2.14]. It seems that NMC has a high selectivity for H_2S and COS over CO_2, but it absorbs the sulfur components only at a moderate rate and its CO_2 activity is very poor indeed.

2.3.2 Chemical Purification Processes

2.3.2.1 Alkanolamine Units

A distinction is normally made between primary, secondary and tertiary amines according to their chemical structure. Primary amines used in practical applications include monoethanolamine (MEA) and diglycolamine (DGA), while among secondary amines diethanolamine (DEA) and diisopropanolamine (DIPA) are widely used. Triethanolamine (TEA) and methyldi-ethanolamine (MDEA) are tertiary amines. The amines used as wash liquors to remove sulfur components and CO_2 from gases differ with respect to their properities and behavior not only between the three above-mentioned groups, but also within the groups by the different substitution groups for the central nitrogen atom. All amines are relatively weak bases which react with the weak acids H_2S and CO_2 forming bonds which can be easily broken by elevated temperatures. Reactions with stronger acids such as thiosulfate or thiocyanate produce so-called heat stable salts, i.e. irreversible degradation products which have to be removed from the wash liquor in order to maintain its efficiency and prevent corrosion. Degradation products are formed also by the reactions between the amines themselves, particularly of MEA, and by reactions with COS and CO_2.

Primary amines, MEA in particular, are chemically stronger than secondary ones, which in turn are stronger than tertiary amines. As their chemical strength decreases, the amines gain in selectivity for H_2S over CO_2. This selectivity is particularly high in the case of those tertiary amines where the nitrogen has no hydrogen atom so that no carbamate can be formed by direct reaction with CO_2.

While CO_2 reacts with primary and secondary amines to form carbamate according to the mechanism

$$CO_2 + R_2NH \rightleftarrows R_2NCOO^- + R_2NH_2^+$$

its reaction with tertiary amines leads to the formation of bicarbonates according to the formula

$$CO_2 + R_2NH + H_2O \rightleftarrows HCO_3^- + R_2NH_2^+$$

H_2S reacts according to the same mechanism as in the case of primary and secondary amines

$$H_2S + R_2NCH_3 \rightleftarrows R_2NCH_3^+ + HS^- .$$

With all amines, COS is removed from the gases by hydrolysis; in the case of MEA, a number of irreversible reactions occur as well, leading to degradation of the wash liquor. Owing to their higher reactivity, primary amines can remove almost 100 % of the COS by a combination of hydrolysis and other reversible or irreversible reactions. The COS removal by the less reactive secondary amines does not exceed something like 75 % [2.15]. The reason for this limited COS removal rate lies in the fact that COS, being the weaker acid, is initially absorbed as effectively as H_2S and CO_2, but is then displaced by the latter two substances as the loading of the wash liquor increases. Numerous gas purification plants can therefore be found in which CO_2 and H_2S are removed to the largest possible extent in a first purification stage (sometimes using wash liquors other than amines) while COS and H_2S are eliminated in a final absorber which normally contains only a few trays and may use DEA as a solvent. An example of such an arrangement is the *Benfield Highpure Process* [2.15].

Table 2.3 summarizes a number of data for primary and secondary amines which are more frequently used in practical applications. It is clear that the theoretical capacity of amine solutions cannot be used to the full extent and that their practical range is limited by their corrosion behavior. The theoretical loading capacity of the amine solutions is limited also by the temperature of the wash liquor. The most beneficial temperature range for amine absorbers is 30–45 °C while the solvent is regenerated by steam stripping at a temperature of 120 °C or more. The regeneration heat demand, i.e. the energy required to heat up the spent solvent to regeneration temperature, to generate enough stripping steam and desorb the impurities, is normally kept to a minimum by providing for intense heat exchange between the cold spent and the hot regenerated solvent. Rising absorption temperatures lead to a smaller absorptivity and – because the dissolution of the acid gases in amine liquors is exothermic so that the liquor is heated up – the increase in temperature associated with the higher loading reduces also the theoretical absorptivity. The dissolution heat of CO_2 and H_2S is almost independent of the concentration of the amine liquor, but decreases steeply as the loading increases; for CO_2 in MDEA, for instance, it amounts to some 60 000 kJ/kmol for a loading of some 0.1 mol CO_2 per mol MDEA and around 33 000 kJ/kmol for a molar ratio of 1. The corresponding figures for H_2S are about 40 000 and 25 000 kJ/kmol H_2S [2.17].

Figure 2.13 shows the layout of a plant designed for high sulfur collection efficiency. H_2S is eliminated from the gas in a two-stage absorber down to a

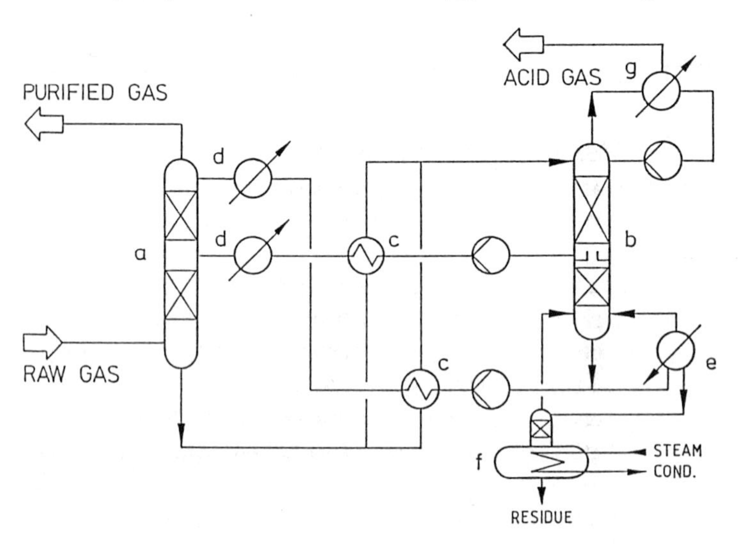

Fig. 2.13. Split flow amine sulfur removal process. (*a*) Absorber; (*b*) regenerator; (*c*) lean/rich solution heat exchanger; (*d*) cooler; (*e*) reboiler; (*f*) reclaimer; (*g*) condenser

Table 2.3. Properties and performance of alkanolamines

	Primary Amines		Secondary Amines	
	MEA	DGA	DEA	DIPA
Molecular weight	61	105	105	133
Reactivity (rel.)	1	2	3	4
Capacity (m^3/m^3)	8 - 30	15 - 50	20 - 70	15 - 55
Corrosivity, design limitations[mol/mol]	0.25-0.45	0,25-0,45	0,50-0,85	0,50-0,85
Stripping req. (MJ/Kmol) (CO_2, H_2S)	87	87	65	66
Reclaiming temp. (°C)	145	190	none	175 (0.1
Reboiler temp. (°C)	120	125	120	120
Heat transfer rate [$KJ/m^2.h$]	3000	1700	2500	2500

residual content in the 0.5 ppm range. The large quantities of sulfur in the raw gas are absorbed by a large quantity of incompletely regenerated wash liquor withdrawn from the central part of the regenerator. Fine purification is then ensured by a smaller rate of completely regenerated wash liquor fed to the absorber top. This slip stream is withdrawn from the regenerator bottom where a reboiler ensures complete regeneration of the wash liquor. The degradation products are eliminated in a reclaimer.

This loop is suitable to reach a residual H_2S content of about 0.6 ppm with some 100 kg of stripping steam per m^3 of solvent, or – if steam saving is the primary objective – to reach a residual content of about 1 ppm with 75 kg of steam per m^3 of amine solution. If the solvent were not split up into a partly regenerated and a completely regenerated slip stream, the same amount of steam would only lead to residual H_2S contents of 2.5 to 5 ppm.

Gas purification processes on the basis of alkanolamines are in some instances offered as standard processes, but are also licensed by various firms as special processes. *Union Carbide* for instance, offers a so-called *Amine Guard Process* which provides for adding an inhibitor to reach high loadings of the amine solution without making it corrosive. The situation is similar for the *SNEA-DEA Process* offered by *Société Nationale Elf Aquitaine*. SHELL *Development Corp.* tailored their *MDEA Process* to selective sulfur removal by an appropriate optimization of the operating parameters while BASF granted licenses for an MDEA variant aiming at maximum CO_2 loading by adding appropriate activators and inhibitors. SHELL offers licenses for an amine wash process termed the *ADIP* Process and using DIPA for selective sulfur removal. This process is frequently used also to eliminate H_2S from *Claus* tail gases.

Unlike all other amine solutions, which have hardly any physical effect, *TEA solutions* are physically effective to a noticeable extent. This property has a favorable effect on the economics of this wash liquor for removing CO_2 insofar as the spent liquor is partially regenerated already by letting its pressure down.

Where the natural selectivity of amine solutions is insufficient to produce a *Claus* gas with the desired sulfur concentration, the sulfur content may be increased by feeding the acid gas from a first stripper to a second absorption stage where the H_2S is absorbed together with only some of the CO_2. The spent liquor from the second absorber is stripped in a second regenerator and then contains the desired high amount of H_2S. Such a system is proposed for instance by *Dow Chemical Cie.* for their *Selectamin Process* [2.18].

New perspectives with respect to capital investment and operating costs of amine units, but also to their H_2S/CO_2 selectivity, are opened up by the so-called *hindered amines* about which more detailed publications first appeared in 1983 [2.19]. Their H_2S/CO_2 selectivity can be used to good advantage especially for the purification of syngas from coal.

This improvement in absorptivity and selectivity results from amine structures which are specially tailored to the required purpose. Some kind of sterically hindered amines overcomes the limits to the absorptivity of primary and secondary amines caused by the fact that the carbamates are highly stable. Carbamate formation is largely suppressed, and the bicarbonate reaction which takes place instead ensures that the loading differential between regenerated and spent liquor can be considerably increased without changing the corrosion behavior of the amine solution.

Another type of hindered amines is characterized by a sharply reduced reactivity with CO_2 while their H_2S absorptivity remains high. The selectivity for H_2S over CO_2, which can thus be achieved, is far better than that of conventional secondary and tertiary amines [2.20].

Hindered amines proved superior also to amines with a conventional structure when used in combination with other chemical or physical solvents, which will be discussed in the following sections.

2.3.2.2 Hot Potash Processes

Hot *Potash Processes* using potassium carbonate or *potash* as it is commonly called, are frequently used to remove CO_2 from syngases. This is due to the fact that the reaction between CO_2 and potassium carbonate taking place in a aqueous phase according to the equation

$$CO_2 + K_2CO_3 + H_2O \rightleftarrows 2\ KHCO_3$$

is favorable for absorption even at elevated temperatures.

This enables CO_2 absorption to be carried out at temperatures close to CO_2 expulsion temperatures in the regenerating stage. The process can thus do without heat exchange between regenerated and spent solvent, which is frequently very complex in the case of chemical purification processes such as the alkanolamine process whose absorbers work at ambient temperatures. As the absorbed CO_2 retains a noticeable vapor pressure even at very small absorption rates and thus is partly expelled when the pressure is let down, and as the desorption heat from hot K_2CO_3 solution is only about 28 000 kJ/kmol (CO_2 dissolution heat in an MEA solution is as high as 37 000 kJ/kmol), only half as much regeneration steam as in a conventional MEA unit is required here.

Although a single-stage hot potash process can remove CO_2 only to residual contents around 0.5 %, this is more than sufficient for a methanol plant. The process can be used with particular advantage wherever the raw gas is saturated with steam at temperatures around 120 °C, for instance downstream of a CO shift conversion unit.

Figure 2.14 shows the very simple layout of a single-stage hot potash unit. As far as desired, CO_2 is removed from the pressurized hot gas in the absorber. The heat input and the absorption heat raises the temperature of the wash liquor above the absorber inlet temperature. Without heat exchange against the regenerated liquor, the spent liquor is flashed into the regenerator, expelling already some part of the dissolved CO_2 together with steam. The remaining CO_2 is then stripped in the regenerator by means of heat from the reboiler. As shown in Fig. 2.14, this heat is in many cases taken from the raw gas before it enters the absorber. The steam leaving the regenerator overhead together with the CO_2 is

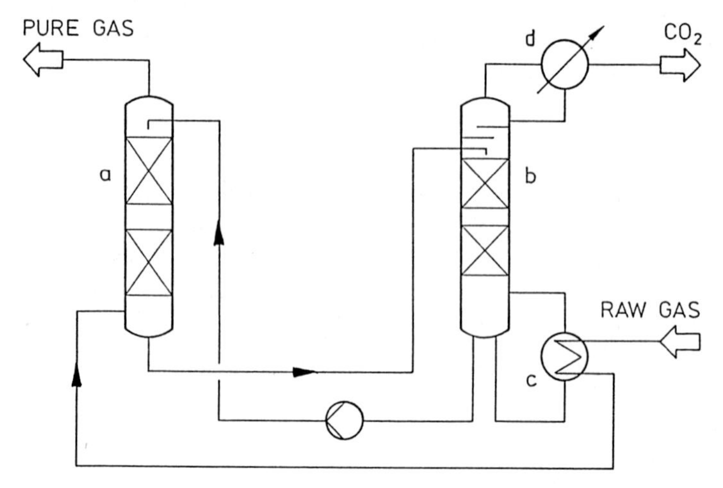

Fig. 2.14. Hot pottash CO_2-removal process. (*a*) Absorber; (*b*) regenerator; (*c*) reboiler; (*d*) condenser

condensed and the condensate is used to wash back sprayed liquor. No reclaimer is required as in the potash unit the solvent cannot degrade.

Various versions of the hot potash process are today offered by a number of different licensors. Apart from measures intended to reduce heat demand, these versions differ by their additives to the potassium carbonate solution, which are in some cases intended to activate the reaction and in others to prevent corrosion.

The *Benfield Process* uses wash liquors containing 25–30 % potassium carbonate and a vanadate additive. As in the case of most hot potash processes, several variants were developed to serve different applications. Split flow systems are used in the absorber and regenerator to remove CO_2 down to a level of 0.05 vol. %; however, in order to remove CO_2 from methanol syngas, the simplest process version consisting of an undivided absorber regenerator loop will do.

Another loop was developed to reduce heat demand. To this end, the spent solvent discharged from the regenerator is flashed to a slight vacuum and the resulting vapors are recycled to the regenerator by means of a steam ejector or vapor compressor.

The *Catacarb Process* uses an additive on the basis of amine borates, whose composition is not disclosed by the licensor.

The *LURGI HP Process* uses borax as a double-acting additive, and a special version based on an appropriate adjustment of operating conditions ensures also a noticeable selectivity for H_2S and COS.

The *Giammarco Vetrocoke Process* uses arsenate as an additive to the potash solution. A modified version of this process has become state of the art as an oxidative purification unit for selective removal of H_2S.

Hot potash processes have been very much superseded by alkanolamine processes. This is partly due to the much greater adaptability of amine solutions to different gas purification jobs, and partly to the fact that the additives used, above all vanadates and arsenates, are hardly compatible any more with stricter and stricter environmental requirements.

2.3.2.3 The Alkazid Process

The two BASF processes, *Alkazid-M* (potassium monomethylamino-propionate) and *Alkazid-DIK* (potassium dimethylaminoacetate), which had been frequently used in the past to remove minute quantities of CO_2 and H_2S or to remove H_2S selectively, have no more chance today of competing with modern coal gas purification processes for methanol production.

The reason lies in the fact that the alkazid solution is highly sensitive to HCN (forming potassium ferrocyanide) and that the sulfur collection efficiency is unsatisfactory, which is essentially due to the poor COS absorptivity of the solvent.

2.3.3 Processes Using Mixed Solvents

The so-called mixed solvents offer a combination of chemical absorption and physical effects. The former binds the sour gas components to the alkalinic wash liquor components (amines) while the latter provides for simultaneous physical absorption by the organic component of the solvent, if the partial pressures are high enough. The major advantage of mixed solvents lies in the fact that a very high sulfur collection efficiency can be reached with reasonable expenditure. The acid gas components are absorbed by the amine in the mixed solvent much in the same way as in the case of aqueous amine solutions, i.e. partly via the gases dissolved in the liquid. Unlike aqueous solutions, however, organic solvents have an absorptivity which is several times higher and the mass transfer to the amine may be much faster. This is essential above all for the absorption of small residual quantities of H_2S and for COS absorption. In addition, the organic component in combination with a certain percentage of water accelerates COS hydrolysis. Consequently, most of the COS is hydrolized to H_2S already in the absorber bottom whereas the major part of the absorber is available for the absorption of H_2S. If, on the other hand, aqueous amine solutions are used, COS hydrolysis proceeds so slowly that it does not only break through in major quantities at the top of the absorber, but that this protraction of H_2S formation towards the absorber top makes it virtually impossible to remove all H_2S as the remaining mass transfer area is no longer sufficient.

Physico-chemical solvents are in certain respects superior also to purely physical COS absorbents. In the case of methanol, for instance, *Henry's* coefficient for COS absorption is only about half that for H_2S. For all practical purposes, this means that the solvent rate has to be doubled if not only H_2S but COS as well has to be eliminated completely (Fig. 2.7). Since, in the case of mixed solvents, COS is converted by hydrolysis to H_2S and this reaction proceeds very rapidly, physico-chemical gas cleaning units are designed only for H_2S removal. Such processes are used to good effect at ambient temperature and pressures of more than 15 bar.

2.3.3.1 The Sulfinol Process

The *SHELL Sulfinol Process* uses Sulfolan (tetrahydrothiophene dioxide) as an organic component and, in its original form, DIPA (di-isopropanolamine) as an alkalinic component, together with some 15 % water [2.21]. Although it had originally been developed for the non-selective removal of acid gas components from natural gases, it has made its way also in the field of syngas cleaning. In addition to CO_2 and H_2S, Sulfinol can remove also major quantities of COS and, unlike aqueous amine solutions, even mercaptanes. Well-designed plants can reach residual sulfur contents of less than 0.5 ppm.

In its configuration, a Sulfinol unit is hardly different from an alkanolamine unit. Unlike the latter, however, Sulfinol tolerates a much higher acid gas loading – approximately twice that of a standard MEA unit – without becoming corrosive. Even though the quantities of degradation products are very small, larger plants are normally equipped with a reclaimer to eliminate the oxazolidone resulting from the reaction of DIPA with CO_2. If major quantities of COS have to be eliminated as well – as it is the case whenever coal with high sulfur contents is gasified at high temperatures – provision is normally made to operate the absorber bottom at slightly increased temperatures, as this will noticeabily accelerate the COS hydrolysis.

Water scrubbing is required both for the clean gas and for the acid gas, not only to prevent carry-over of DIPA and sulfolan to downstream units, but also because of the very high sulfolan cost. No distillation section is required. Since both DIPA and sulfolan have higher boiling points than water, the water entrained with the feed gas or used in the secondary scrubbers can be discharged from the unit together with the acid gas.

As the absorptivity of the solvent is high, little steam is required to regenerate it. The regeneration stage – sometimes preceded by a flash stage – operates at approximately 125–130 °C. For example, if the sulfur components are to be removed to less than 0.5 ppm from a 50 bar gas containing 4.3 % by vol. of CO_2, 0.3 % of H_2S and 0.01 % COS, some 150 m^3 of solvent are needed per 100 000 m^3 of raw gas. Since mass transfer is very rapid, only about 40 trays are needed in the absorber. Regeneration of the spent solvent requires some 3.6 kg of steam of not less than 3.5 bar per m^3 of acid gas. Considering that the CO_2 is also removed more or less completely, this means that about 53 kg of steam are required per m^3 of sulfur components removed. The acid gas leaving the regeneration section has a sulfur content (H_2S + COS) of only about 6.7 % by volume and is therefore not suitable for further treatment in a *Claus* unit operating with air (see Sect. 5.2).

A combined effort at both reducing steam demand and obtaining an acid gas with higher sulfur contents led to the development of a *selective* version of the Sulfinol process which had the DIPA replaced by MDEA. However, with this M-Sulfinol process, as it is called, good selectivity is in this case achieved at the expense of higher residual sulfur contents in the clean gas. If the residual H_2S content is required to be less than 10 ppm, much of the selectivity is lost, since for instance for 1 ppm residual H_2S the wash liquor loop becomes so large that most of the CO_2 in the gas is absorbed as well. In contrast to the non-selective process version, only about half of the CO_2 would be removed from the raw gas described in the above example, but a residual sulfur content of more than 10 ppm would then have to be tolerated. The solvent demand per 100 000 m^3 of raw gas decreases considerably, and only about 25 kg of regenerating steam are required per m^3 of H_2S + COS. The acid gas contains some 15 % by vol. of sulfur and can now be processed in an air-blown *Claus* plant.

If this process version is used to desulfurize raw gases for methanol production, a simple adsorption stage using zinc oxide is required for additional fining. In some cases the gas cleaning unit can even do without such a fining stage because the small residual quantities of sulfur – sometimes after the COS has been converted to H_2S in a CO-shift conversion stage – can be removed in a CO_2 scrubber together with the surplus CO_2.

2.3.3.2 The Amisol Process

The Amisol process was developed by LURGI on the basis of methanol with its high physical absorptivity, which had already proved its merits in the Rectisol process. The standard version – designed as a non-selective acid gas removal unit – used MEA or DEA as chemical components. As in the Sulfinol process, a small percentage of water – approximately 12 % by wt. – accelerates COS hydrolysis. By cooperation of chemical and physical components, the Amisol process reaches residual sulfur contents of less than 0.1 ppm at medium pressures of 10 to 50 bar and at ambient temperature in the absorption section. Although a single stage will do the job, both the absorber and the regenerator are normally operated in a split-flow system in order to reach residual sulfur contents which are acceptable for methanol synthesis and at the same time keep the heat demand for regeneration down. The major part of the acid gas components are absorbed by a larger solvent stream which has been only moderately regenerated while a small and thoroughly regenerated slip stream is used for fining.

As the solubility of hydrogen and carbon monoxide is practically independent of the prevailing temperature (see Fig. 2.7), the rate at which these two gas components are dissolved at ambient temperature economically justifies an interim flash stage followed by H_2 and CO recompression, above all for large plants and high pressures.

In addition to its high absorptivity for acid gas components, methanol as the main component of the Aminsol solution is characterized by a very low viscosity, which favors mass transfer and reduces the absorber heights, and has a low boiling point. This latter property ensures that the spent liquor can be regenerated at about 90 °C so that cheap waste heat can be used, which is normally abundant in coal-based methanol plants. Table 2.4 contrasts some data for the selective and non-selective versions of the Sulfinol and Amisol processes. It is evident at first glance that the selective versions require a great deal less steam for regeneration.

Figure 2.15 shows the process flow diagram for a single-stage Amisol unit. Having been cleaned in the absorber at something like 40 bar, the gas has to be fed to a water scrubber to avoid major solvent losses which would otherwise be

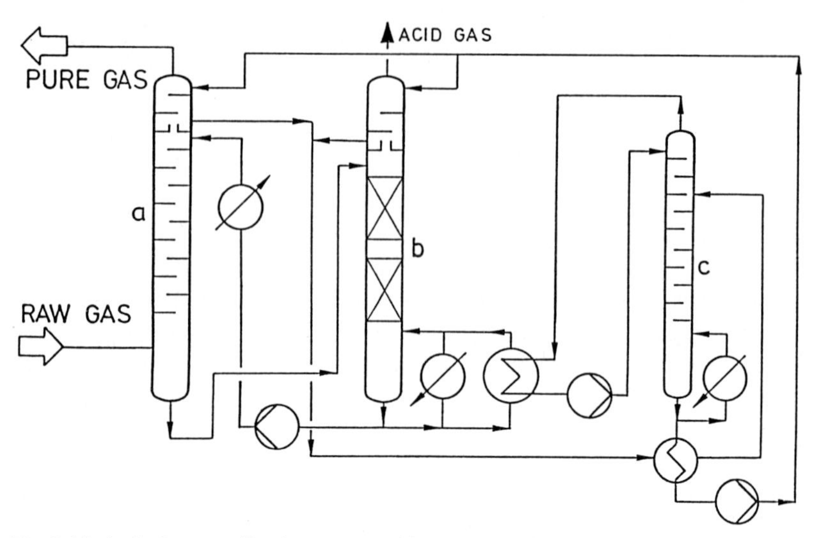

Fig. 2.15. Amisol gas purification process. (*a*) Absorber; (*b*) regenerator; (*c*) methanol/ water distillation

incurred because of the high methanol vapor pressure at ambient temerature. The spent solvent is flashed at approximately 10 bar and the flash gas recompressed into the absorber. The solvent is then regenerated with heat in the regenerator. Heat exchange between hot regenerated and cold spent solvent is normally not reasonable under technical aspects since already a moderate acid gas load heats up the solvent to a level at which its temperature differs by only a few degrees from that of the regenerator bottom temperature. The acid gas expelled in the regenerator, too, is fed to a water scrubber.

The scrubber effluent, which contains some methanol, is fed to a distillation column to expel the methanol overhead and obtain clean water at the bottom. This distillation column operates at a slightly elevated pressure to ensure that the overhead vapors are condensed in a separate reboiler at the regeneration column and can thus provide part of the heat required by that column. The remaining regeneration heat demand may be covered by waste heat at temperatures above 90 °C.

More selective versions were developed for the Amisol process, too. One of them uses diethylamine (DETA) as a chemical component. This alkylamine component is characterized by such a high COS absorptivity that no water needs to be added to the solvent to accelerate COS hydrolysis. Unlike MEA and MDEA, its stronger physical effect when dissolved in methanol makes it more selective for COS over CO_2. Table 2.4 illustrates the high natural selectivity of this mixed solvent. DETA differs from a wash liquor made up of DEA and methanol also insofar that it removes any mercaptanes that may be contained in the raw gas at least partially.

An Amisol version which is suited particularly to remove CO_2 uses methyldiethanolamine (MDEA) as a chemical component. Owing to its excellent physical

Table 2.4. Performance of mixed solvents

Chem. Component	Sulfinol DIPA	Sulfinol MDEA	Amisol DEA	Amisol DETA
Raw gas CO_2 (% vol.)		4.30		
H_2S (% vol.)		0.30		
COS (% vol.)		0.01		
Pure gas CO_2 (% vol.)	0.1	3.2	0.1	2.1
H_2S COS } ppm	0.5	12	0.1	0.1
Mercaptane removal	yes	yes	no	partly
H_2S+COS in acid gas (% vol.)	6.7	20.0	6.7	13.0
Solvent required (m^3/100 000 m^3 raw gas)	150	70	180	75
Electricity required (kWh/100 000 m^3 raw gas)	340	130	320	135
Cold required (-40°C) (GJ/100 000 m^3 raw gas)	-	-	-	-
Reg. Steam required (kg/100 000 m^3 raw gas)	16400*	6500*	16900**	7500**
(kg/m^3 H_2S + COS)	53.0	21.0	54.5	24.2

* steam at 3.5 bar min.
** steam of waste heat at 90°C min.

effect, the combination of this tertiary amine with methanol is particularly suitable for the removal of large CO_2 quantities. Whenever the syngas requires to be only moderately clean of CO_2, as this is the case for methanol synthesis, the solvent can be regenerated largely by flashing. In comparison with other CO_2 absorption processes, only very little additional steam is required for regeneration.

2.3.3.3 Oxidative Gas Cleaning Processes

As already mentioned at the beginning of the chapter on gas cleaning, the oxidative processes do not eliminate the acid gas component by a reversible method and release it again as the solvent is regenerated. Rather, a reaction takes place in the wash liquor so that the absorbed gas component is obtained in some other form. This is of particular interest for selective desulfurization of raw gases and for H_2S removal from acid gases to facilitate sulfur recovery on the one hand and, on the other, to bring down the residual sulfur content of the CO_2 to acceptable limits before it is discharged to the atmosphere.

In this sector, the *Stretford Process* offered by *British Gas Corp.* and the *Takahax Process* of *Tokyo Gas Cie.* Ltd. have prevailed over the older processes of this type such as Giammarco-Vetrocoke, Ferrox, Thylox and others [2.4]. A certain preference for the former can be observed in the USA and Europe, whereas the latter is more frequently used in Japan. The Takahax process, which exists in 4 versions in combination with the Hirohax process developed by *Nippon Steel*, is specifically tailored to coke oven gas cleaning and may therefore be left unconsidered here.

The first stage of the Stretford process absorbs H_2S over sodium carbonate contained in the wash liquor together with sodium ammonium vanadate, anhydrous citric acid and anthraquinone disulfonic acid. The reaction

$$Na_2CO_3 + H_2S \rightarrow NaHCO_3 + NaHS$$

produces sodium hydrogen carbonate and sodium hydrogen sulfide. The HS-ion is oxidized by the vanadate in the solution partly already in the scrubbing section and completely in the reaction section:

$$HS^- + 2V^{5+} + OH \rightarrow H_2O + 2V^{4+} + S.$$

The V^{4+} ions are re-oxidized to vanadium pentoxide by adding air in the oxidation section, the anthraquinone disulfonic acid acting as oxygen carrier and accelerator.

Figure 2.16 illustrates the process in the form of a flow diagram. When the H_2S has been absorbed from the acid gas, the spent liquor, so to speak, from the absorber is fed to the reaction tank where the remaining H_2S is oxidized to elemental sulfur. The solution containing the suspended sulfur flows to the oxidizer where the semiquinone is re-oxidized to anthraquinone. A blower delivers air to provide the necessary oxygen. This air causes the sulfur to float up into a supernatant foam, which can then be skimmed off into the sulfur slurry tank. The "regenerated" solution flows to the pump surge tank to be pumped back to the absorber.

The sulfur recovery unit serves to separate the sulfur from the entrained solvent before it is fed to the sulfur smelter. The solvent may be removed either in a drum filter, in which it is washed at the same time, or in several centrifuge stages with interim wash stages. The molten sulfur may be withdrawn continuously; the recovered solvent is recycled to the absorption loop. The sulfur is normally 99.9 % by wt. clean, containing only traces of contaminants such as vanadium. Salts such as sodium thiosulfate and sodium thiocycanate, which may be produced by secondary reactions, have to be eliminated from the system to prevent their build-up and eventual crystallization.

If the gas to be cleaned contains significant quantities of HCN, it will be advisable to remove them before the gas enters the desulfurization stage. To this end, the HCN is normally reacted with sodium polysulfide in a pre-wash stage to produce sodium thiocyanate.

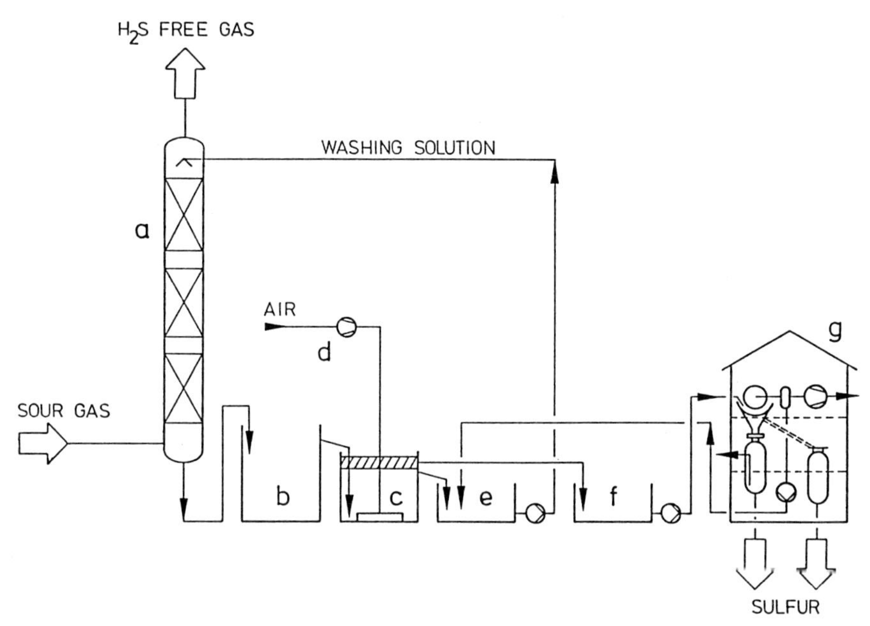

Fig. 2.16. Stretford H_2S removal process. (*a*) Absorber; (*b*) reaction tank; (*c*) oxidizer; (*d*) air blower; (*e*) pumping tank; (*f*) sulfur slurry tank; (*g*) sulfur recovery

The Stretford process is suitable to remove H_2S contents in the clean gas to something like 1 ppm even if H_2S partial pressures are low and CO_2 contents in the acid gas high. The process is therefore used frequently to remove H_2S from atmospheric offgas flows. The Stretford process does not eliminate any COS that may be contained in the gas to be cleaned [2.22].

2.3.3.4 Pre-Wash Stages

It has already been mentioned occasionally in the preceding chapters, that in addition to the acid gases CO_2 and H_2S, which normally occur in higher concentrations and can be satisfactorily removed from raw gases by means of the gas cleaning processes described above, there are a number of pollutants which occur at low concentrations and may have to be either converted or hydrogenated or hydrolized, for instance COS and organic sulfur compounds or other substances which have to be absorbed in pre-wash stages. In coal gases, the latter include such high-boiling organic compounds as paraffines and olefins, on the one hand, and acids like HCN and formic acid, as well as other inorganic compounds such as ammonia, on the other.

The kind of such pollutant traces and their combinations are decisive for the selection of the pre-wash stage. HCN, ammonia, formic acid and other substances having a sufficiently high *Bunsen* absorption coefficient can be effectively removed from the gas already by simple *water scrubbers*. Since good absorption always goes along with difficult desorption, large amounts of stripping fluid are usually required to regenerate the wash liquor. Air can in many cases be used

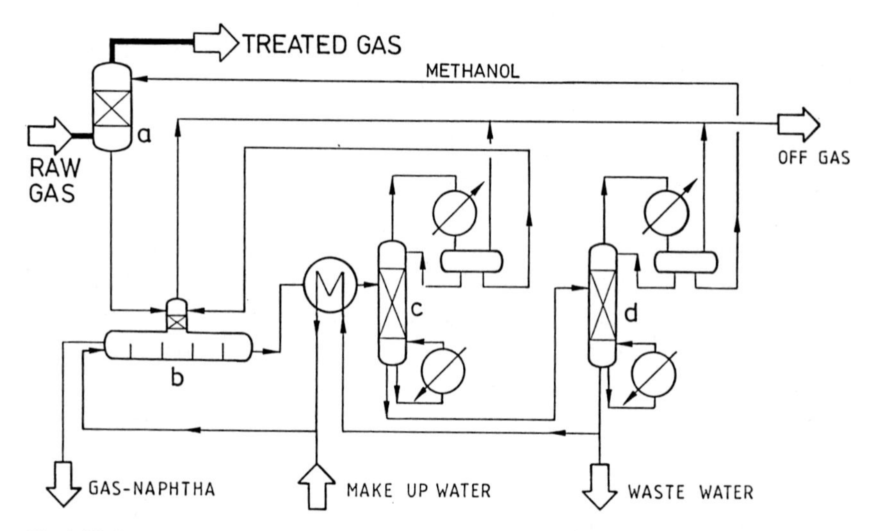

Fig. 2.17. Rectisol prewash unit. (*a*) Pre absorber; (*b*) extraction; (*c*) and (*d*) distillation

as a stripping fluid, for instance whenever the system includes a Claus unit, and this air can then be used for combustion together with all the stripped pollutants. Alternatively, steam may be used for stripping, but this steam has then to be condensed to separate it from the vapours, and the pollutants will eventually have to be burnt.

Hydrocarbons may be removed in *oil scrubbers*. However, the natural vapour pressure of these solvents sometimes requires the use of higher boiling oil fractions in the pre-absorber, which can then no longer be boiled out indirectly with steam in a conventional heat regenerator, but can be heated to the boiling point only by means of live steam or in fired reboilers.

The Rectisol process can be used for a virtually universal pre-wash stage [2.4]. A small quantity of cold methanol will suffice to clean coal gases from medium-temperature pressure gasification sufficiently to keep all those pollutants away from the main scrubber which are either undesirable in higher concentrations in the acid gas, for instance ammonia, or which should be kept away from the equipment in order to avoid the use of expensive materials, for instance HCN and formic acid, or else which must be expected to break through into the clean gas because they are difficult to desorb, for instance hydrocarbons.

Figure 2.17 shows the arrangement of such a Rectisol unit with a pre-absorber in which the above-mentioned pollutants are washed out by methanol. The spent methanol is regenrated via extraction with water and subsequent two-stage distillation. The extraction stage recovers the hydrocarbons and the distillation stages remove the other pollutants.

2.3.4 COS Shift to Assist Gas Purification

Like almost all gasification processes, coal gasification produces not only H_2S but also COS from the sulfur contained in the raw material. For instance, coal containing 1.3 wt. % of sulfur which is fed to a high-temperature gasification unit in the form of a coal/water slurry leads to approximately 0.6 vol. % of H_2S and 0.04 vol. % of COS in the raw gas.

Wherever the absorption or adsorption processes used to clean the gases have a high absorptivity for H_2S but can eliminate COS only at high cost or not at all, the processes are designed in such a way that the sulfur contained in the gas reaches the gas purification unit in the form of H_2S. Since, however, most modern gas purification units are capable of removing not only H_2S but also COS and other organic sulfur components efficiently enough to meet the requirements for methanol production, COS hydrolysis is today used only in special cases.

2.3.4.1 Fundamentals of COS Shift Conversion

A distinction is made between COS hydrogenation converting the COS to H_2S at temperatures between 350 and 450 °C together with CO shift conversion, and COS hydrolysis where the COS is converted in a separate conditioning stage at temperatures between 120 and 300 °C. The former produces an equilibrium according to the formula

$$H_2 + COS \rightleftarrows H_2S + CO.$$

Table 2.5 lists some k_p figures for this reaction. The COS hydrogenation equilibrium in the temperature range used for high-temperature conversion of carbon monoxide is so favorable that, for the above-mentioned gas, the original COS/H_2S ratio of 1 : 15 diminishes to 1 : 200 downstream of the CO shift conversion stage, or – in other words – that the COS content decreases from 400 to 20 ppm.

Since in almost all methanol production plants only some part of the raw gas is carried through a CO shift conversion unit, COS hydrolysis is more important for coal gases than COS hydrogenation which takes place parallel with the water gas reaction. The equilibrium equation for COS hydrolysis is

$$COS + H_2O \rightleftarrows H_2S + CO_2.$$

Some equilibrium figures (k_{p2}) for this reaction are also listed in Table 2.5.

Table 2.5. Equilibrium constants for COS-hydrogenation (K_{p1}) and COS-hydroloysis (K_{p2})

°C	$K_{p1} = \frac{H_2S \cdot CO}{COS \cdot H_2}$	$K_{p2} = \frac{H_2S \cdot CO_2}{COS \cdot H_2O}$
100	2.2	$3.16 \cdot 10^4$
200	6.5	$2.75 \cdot 10^3$
300	10.0	$5.55 \cdot 10^2$
400	12.9	$1.85 \cdot 10^2$
500	15.4	81.0
600	17.3	43.1
700	18.6	26.1
800	19.7	17.3

2.3.4.2 *COS Conversion Catalysts*

COS hydrogenation at high temperatures occurs with satisfactory space velocities over both iron oxide and cobalt/molybdenum catalysts as they are used for CO shift conversion.

For decades, chromium/aluminium and copper/chromium/ aluminium catalysts have been used at temperatures between 250 and 320 °C to achieve a satisfactory approximation to equilibrium at space velocities between 2000 and 4000 m^3/m^3h. In recent years, pure aluminium catalysts, and very recently also platinum catalysts whose activity for COS (and CS_2) conversion is good already at temperatures between 120 and 200 °C have won favor for COS hydrolysis. Such low-temperature operation is a great advantage as the steam rate required for equilibrium decreases steeply with decreasing reaction temperature.

These catalysts are completely inactive with respect to the water gas reaction and an aluminium catalyst can successfully be used to reduce the COS content of the above-mentioned raw gas to a few ppm by hydrolysis at a temperature of 200 °C.

2.3.4.3 *COS Hydrolysis Unit*

Figure 2.18 illustrates the extremely simple layout of a COS hydrolysis unit. Depending on the requirements, the steam content of the raw gas is appropriately adjusted by quenching the gas with water (direct cooling) – normally still in the coal gasification unit – or by adding steam to it after it has been preheated to the required inlet temperature for COS conversion. Let us assume that the COS content of the gas referred to in the previous chapter, containing some 3 vol. % of CO_2 in addition to 0.6 vol. % H_2S and 0.04 vol. % COS, is to be reduced to 2 ppm. The gas is assumed to be available at a pressure of 30 bar and the reaction is to take place at 200 °C. A water content of 0.033 mol./mol. of raw gas can be derived from the k_p figure, corresponding to a saturation temperature of 90 °C.

The raw gas is preheated to 200 °C in a preheater and steam is added as required before the mix is passed over the hydrolysis catalyst. As the reaction

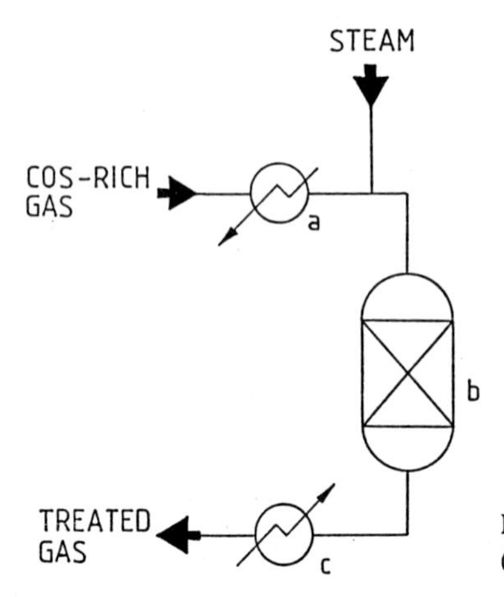

Fig. 2.18. COS hydrolysis unit. (*a*) Preheater; (*b*) reactor; (*c*) cooler

involves virtually no change in the overall enthalpy, the converted gas leaves the reactor without having changed its temperature.

2.4 Conditioning of Methanol Synthesis Gas

2.4.1 Conditioning to Adjust the Stoichiometric Number

Almost without exception, the raw gases produced from coal contain more carbon monoxide than required for methanol production. This is why most coal-to-methanol plants include facilities to shift some of the carbon monoxide contained in the raw gas.

A distinction is normally made between clean gas conversion where the CO shift conversion unit is preceded by a gas purification stage to remove the major part of the sulfur components and any higher hydrocarbons that may be present in the gas, and sour gas conversion which accepts the untreated coal gas as a feed. A special type of the latter is what has been termed raw *gas conversion*, where considerable quantities of high boiling hydrocarbons are not eliminated from the raw gas but are left to pass the CO shift conversion stage.

Depending on the temperatures at which the carbon monoxide is shifted, another distinction is made between high-temperature shift conversion (300–500 °C) and low-temperature shift conversion (180–280 °C). Low- temperature shift conversion is, however, normally used only if the residual CO content in the converted gas has to be very low. As this is not the case for methanol production, and as there is no reason to put up with the high vulnerability to sulfur of the copper catalysts used for low-temperature conversion nor their considerable cost, the following description will be limited to high-temperature conversion.

2.4.1.1 Fundamentals of CO Conversion

Independently of the overall pressure, but depending upon the temperature of the gas mix, there is a firm relationship – defined by the water gas equilibrium – between carbon monoxide, carbon dioxide, hydrogen and steam.

The CO shift reaction described by the equation

$$CO + H_2O \rightleftarrows H_2 + CO_2$$

is exothermic with an enthalpy change of H = −42 700 kJ/kmol. The equilibrium constant is defined by

$$k_p = \frac{p_{CO} \cdot p_{H_2O}}{p_{H_2} \cdot p_{CO_2}}.$$

The literature contains numerous formulations for this dependence on temperature; one frequently used equation [2.23] is

$$\lg k_p = -2\,059/T + 1.5905 \cdot \lg T - 1.817 \cdot 10^{-3} \cdot T + 5.65 \cdot 10^{-7} \cdot T^2$$
$$-8.24 \cdot 10^{-11} \cdot T^3 - 1.5313.$$

Although the heat of reaction is also dependent on temperature, the widely used calculation formula

$$Q_R = 10\,681 - 1.44 \cdot T - 0.4 \cdot 10^{-4} \cdot T^2 - 0.084 \cdot 10^{-6} \cdot T^3$$

leads to only slight differences over the technically interesting temperature range. For high temperatures around 500 °C, Q_R is something like 39 800 kJ/kmol. For the high-temperature conversion process, the interesting temperatures at which CO is shifted at an industrial scale range between 300 and 530 °C. The upper limit is defined by the thermal resistance of the catalysts used and the lower limit by their decreasing activity.

In all practical applications, the reaction temperature is found to differ from the calculated equilibrium temperature so that the ideal equilibrium is never reached. This deviation for the high-temperature range is approximately 20 °C at a temperature of 400 °C and only 10 °C at a temperature of 520 °C.

2.4.1.2 CO Conversion Catalysts

The CO shift conversion of gases containing only small quantities of impurities such as sulfur or condensable hydrocarbons, i.e. the so-called clean gas conversion process, uses iron oxide catalysts doped with other elements such as chromium and aluminium as reaction accelerators and temperature stabilizers in the high-temperature range. Small sulfur contents may often come also from binders or from the pressing tool lubricants. The catalysts used today are exclusively of pressed type, marketed in the form of tablets or cylinders whose mechanical strength is considerably higher than that of the formerly used crushed catalysts (brown oxide). Under otherwise identical conditions, the smooth surface and regular shape also lead to a smaller pressure drop of the gas mix flowing through the catalyst bed.

High-temperature catalysts are supplied in the oxidized condition, i.e. containing iron in the form of Fe_2O_3 and chromium in the form of Cr_2O_3. The catalysts are not normally activated by reduction until they have been filled into the vessel, because the reducing process makes them lose as much as 30 % of their strength. The catalysts are therefore reduced in situ, using hydrogen–containing gases – mostly the clean gas itself – but also mixtures of superheated steam and nitrogen as reducing agents.

Iron oxide catalysts can well tolerate minor quantities of impurities. Detrimental are fluctuating sulfur concentrations which will frequently sulfurize and desulfurize the catalyst grain and thus impair its mechanical stability. Rapid changes in temperature, as they may occur when the hot catalyst gets into contact with entrained water should also be avoided. They, too, will lead to disintegration of the granular structure and will in most cases lead to a steep increase in pressure drop due to the formation of dust pockets. All heating and cooling operations or pressure changes should be made gradually.

Whenever the gases processed by CO shift conversion still contain major quantities of sulfur or high-boiling hydrocarbons such as tars (in the case of raw gas conversion), the iron oxide catalysts are replaced by catalysts containing essentially cobalt and molybdenum. Their ranges of application and activity are comparable to those of iron oxide catalysts, but the activity of cobalt/molybdenum catalysts can still be noticeably increased over the pressure range from 40–80 bar, whereas in the case of iron catalysts an increase in operating pressure beyond 40 bar has virtually no influence on the required catalyst volume. Cobalt-molybdenum catalysts reach their full activity only in the sulfurized condition. Consequently, this catalyst has to be sulfurized before or when the unit is commissioned, and the H_2S/H_2O ratio has to be consistently kept above 1/1 000 in order to maintain the catalyst activity.

The catalyst volumes required for a particular application depend on the partial pressures of the reaction components, on the required CO content of the converted gas, and on the reaction temperature. The operating parameters and the split-up into several process stages are normally selected so as to ensure volume flow rates of 1 000 to 3 000 m^3 gas/m^3 of catalyst and hour in the high-temperature range.

2.4.1.3 Clean Gas Conversion

Clean or precleaned gases may be available either dry, i.e. containing almost no steam at ambient temperature, or with high steam contents at temperatures up to 400 °C. While CO shift conversion units for cold gases normally include a so-called saturator/desaturator system to keep the high-pressure steam demand down, hot gases containing in most cases all the steam required for the CO shift conversion reaction are normally reacted according to the hot conversion principle.

Saturator/desaturator systems go far to improve the economics of CO shift conversion, provided that the gases are available in a cold condition. If the CO content of a gas containing some 45 % CO, 45 % H_2 and 5 % CO_2 is to be reduced to 3 %, the K_p figure of 0.114 calculated for 430 °C (410 °C operating temperature plus 20 °C deviation from equilibrium) leads to a steam demand of approximately 1.47 mol per mol of gas, consisting of about 1.06 mol of equilibrium steam plus 0.41 mol of conversion steam. Hence, for a shift conversion unit operating at for instance 60 bar, almost 3 kg of steam at a pressure of not less than 63 bar would have to be provided per m^3 of carbon monoxide to be shifted.

Figure 2.19 illustrates the operating principle of a saturator/desaturator system in conjunction with a two-stage CO shift conversion unit. Cold gas enters the saturator which in this case is operated at 60 bar. As it moves up through the packing, the gas is contacted with hot water of 250 °C so that it is saturated with steam. On leaving the saturator, the gas contains 0.85 kg water per m^3. The additional steam rate of 0.33 kg/m^3 required for CO shift conversion is added to the saturated gas and the gas/steam mix is then heated to approximately 360 °C and fed to the first reaction stage. As the shift reaction is exothermic, the temperature of the mix increases to some 500 °C until equilibrium is reached at about 7.9 % CO. The hot gas is then used first in a gas/gas heat exchanger to preheat the feed mix and then in a gas/water heat exchanger to preheat the circulating water. Having cooled down to 385 °C, it enters the second reaction zone where

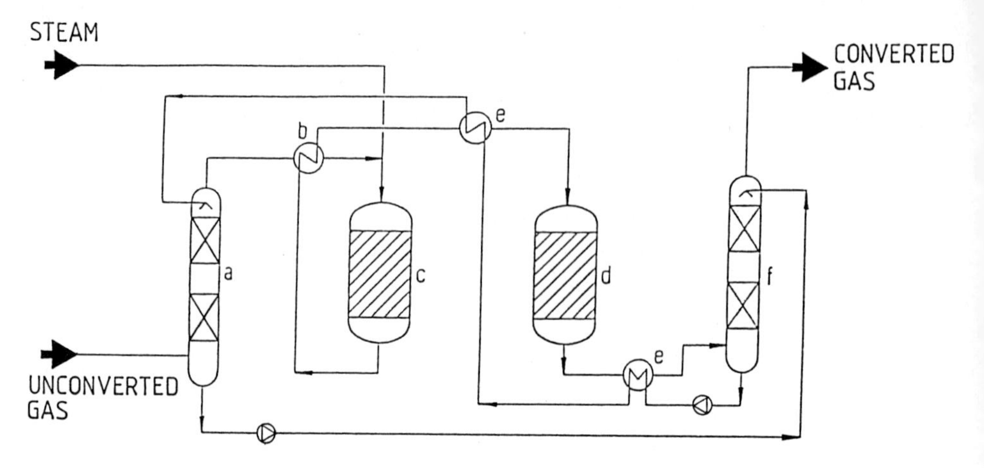

Fig. 2.19. CO-shift conversion process. (*a*) Saturator; (*b*) heat exchanger; (*c*) and (*d*) reactor; (*e*) water preheater; (*f*) desaturator

the CO is further converted to a residual content of approximately 3 % and the temperature increases to 410 °C.

Having left the second reaction stage with a residual steam content of 0.6 kg/m^3, the converted gas passes a second gas/water heat exchanger to heat up circulating water before it enters the desaturator at approximately 235 °C. As it rises through the desaturator packing, it is contacted with colder water trickling down so that the major part of the steam is condensed and the aqueous effluent is heated to a temperature about 2 °C below the dew point of the converted gas (approximately 220 °C).

The water from the desaturator is heated to something like 250 °C in the two heat exchangers and fed to the saturator. It leaves the saturator again at approximately 190 °C at a rate that has been reduced by the quantity of steam fed to the gas. As the additional steam introduces heat to the system, the gas temperature of approximately 200 °C at the desaturator outlet is considerably higher than the saturator inlet temperature. This surplus heat may be used, for example, to preheat boiler feedwater in a converted gas/BFW heat exchanger. Normally, however, the desaturator is subdivided into two stages, the lower of them feeding directly on the saturator effluent, while a small slip stream is used for heat exchange and then fed cold at the desaturator head.

Wherever the catalyst volume required for CO shift conversion permits doing so – for the above-mentioned gas this is possible up to a throughput of approximately 50 000 m^3/h – the two series-connected reaction stages are often replaced by the so-called *countercurrent cooling reactor* as shown in Fig. 2.20.

Whereas the reaction in a simple shaft reactor is adiabatic and therefore requires a two-stage operation, the temperature profile along a countercurrent reactor is much more favorable. On entering the ring space between the reactor shell and the catalyst vessel at approximately 230 °C, the gas/steam mix flows downwards and then enters the tubes in the catalyst vessel. As it flows upward through these catalyst surrounded tubes, the mix is heated up to approximately

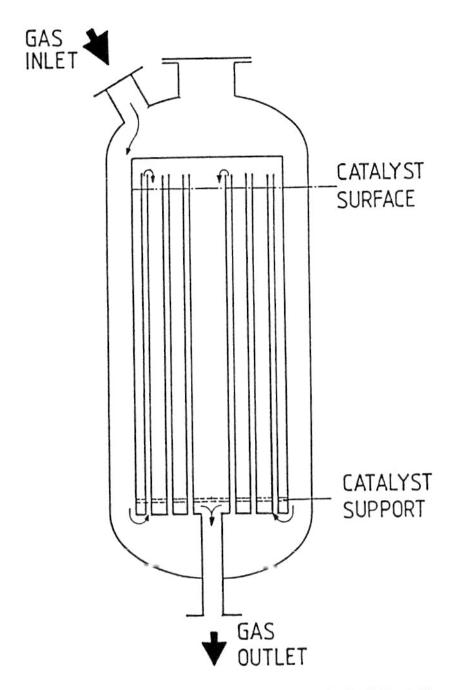

Fig. 2.20. Countercurrent gas cooled CO-shift converter

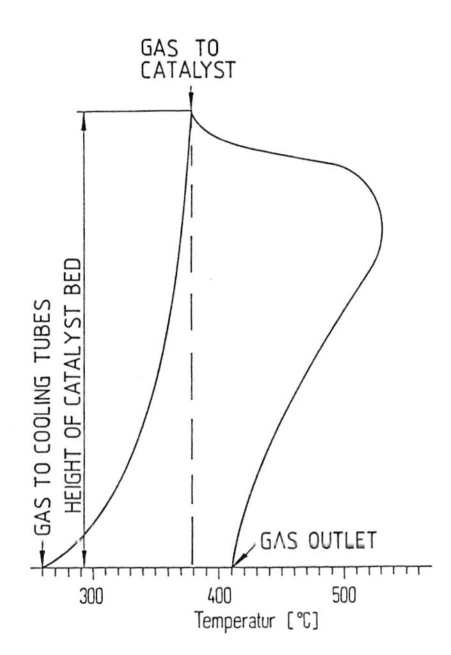

Fig. 2.21. Temperature profile of a countercurrent reactors

380 °C and then enters the catalyst bed from the top. The interaction between the increase in temperature due to the exothermic reaction and the cooling effect produced by the gas mix flowing through the cooling tubes results in the temperature curve as shown in Fig. 2.21. This type of reactor combines a high reaction rate at high temperature in the top section with a favorable low-temperature equilibrium in the bottom section. As the catalyst activity decreases faster at high temperatures, this design offers also some advantages with respect to the desirable long-term activity of the catalyst at those points which are important to establish the equilibrium, i.e. at the reactor outlet.

Much easier is the design of the CO shift conversion unit for *hot gases*, which are obtained from coal gasification units already with a high steam content and relatively high temperatures of 300–400 °C. These gases are converted in one or two reaction stages, if necessary after they have been preheated against hot converted gas and some steam has been added. The heat that needs to be removed between the first and second reaction stages in a two-stage CO shift conversion unit is normally used to preheat the feed gas and sometimes also boiler feed water. The temperature is reduced at this point occasionally also by quenching with water. This has the effect that not only the temperature is adjusted to the desired level for the second reaction stage, but that the steam content in the partly converted gas is increased as well, with an ensuing improvement in the residual CO content.

An optimum H_2/CO ratio in the methanol syngas can in all cases be ensured by a single-stage CO shift conversion unit. However, it will often be more cost-effective to convert only part of the clean gas to CO contents around 3–4 vol. %, while the remaining gas flow bypasses this section (and perhaps also the downstream CO_2 scrubber).

2.4.1.4 Sour Gas Shift

The term *sour gas shift* is used whenever a CO shift reaction is performed on gases from which the sulfur components had not been previously removed. Except for the catalysts, this sour gas shift is not much different from clean gas shift conversion, but unlike the latter, it is more frequently used in its hot than in its cold form. The reason is that in numerous coal gasification processes it is preferable to quench the hot raw gas with water rather than using waste heat recovery equipment at high temperature levels (such as waste heat boilers). As quenching is always associated with the elimination of solids (ash, coal dust) from the raw gas, it also avoids choking of boiler tubes and deposits on the conversion catalyst. Only in rare cases is the sulfur content of the coal so high that there is a risk of sulfide corrosion at high temperature ranges; hence, the materials are also very much the same as for clean gas conversion, and precautions against CO_2 corrosion can be limited to the temperature range below the steam dew point. A less frequently used type of CO conversion in hot gases is what has been termed *quench conversion*. Whereas shift conversion in the temperature range that has been considered so far requires catalysts, the quench variant takes advantage of the fact that the water gas reaction takes place even without catalysts above approximately 950 °C with well-acceptable retention times.

A gas produced by high-temperature gasification of a coal slurry [2.24], for instance, and containing some 10 % CO_2, 45 % CO and 36 % H_2 (in addition to sulfur components and a few inerts) could be shifted without any catalysts simply by quenching with water to obtain a syngas which would be appropriate for methanol production. To this end, 0.55 kg of water would have to be added to the raw gas containing some 0.4 kg of steam per m^3 of gas at approximately 1 500 °C. The temperature of the mix would in this way be adjusted to approximately 950 °C and about one fifth of the CO in the raw gas would be converted. The converted gas would then contain 32 % CO_2, 19.5 % CO and 40 % H_2 so that – after the CO_2 has been washed out to approximately 3 % – the methanol synthesis gas will have a stoichiometric ratio of 2.07.

The term *raw gas conversion* is used for CO shift conversion of gases containing not only sulfur but also major quantities of condensable hydrocarbons, and possibly even solids. This is the case for most medium-temperature coal gasification processes, for instance, the LURGI pressure gasification process. The raw gas from such gasification units is normally quenched more or less to the dew point at the gasifier outlet, so that solids and the major part of the high-boiling hydrocarbons are eliminated. Any solids remaining in the gas as well as the condensed hydrocarbons (droplets) have to be thoroughly removed by mechanical means before the gas enters the CO shift conversion unit in order that deposits on the catalyst are kept to a level at which a satisfactory catalyst service life is still ensured.

The cobalt/molybdenum catalyst used for this application produces an extremely desirable side effect. It hydrogenates the unsaturated hydrocarbons in the raw gas almost completely to saturated components.

2.4.2 Conditioning to Improve Syngas Yield

As described in Chap. 3 below, complete conversion of the reacting gases in the methanol synthesis section leads to the production of two different types of residual gases. The major quantity, termed purge gas, is withdrawn from the synthesis loop at pressures which are only slightly below the pressure in the synthesis unit. The gases dissolved in the crude methanol, on the other hand, are obtained at lower pressures, partly at the preliminary methanol flushing stage and partly in the low-boilers column of the methanol distillation section. In the latter case, the gases are obtained at atmospheric pressure and normally consist of more than 50 % CO_2; they account for only about 0.5 % of the syngas feed rate. It is therefore uneconomic to process these flush gases and they are incinerated instead. Depending on the syngas methane content, which may fluctuate between 0.5 and 12 vol. % depending on the gasification process used, the purge gas rate may account for as much as 20 % of the syngas feed rate. If methane contents in the syngas are high, the purge gas may contain 30–40 % methane and it will be of vital importance for the economics of the process as a whole to condition this gas and recycle it into the process. Doing without conditioning and recycling the gas may be worthwhile only in exceptional cases, for instance if major quantities of sulfur-free fuel gas are required or if the purge gas can be methanated to SNG which is then fed into a pipeline. The purge gas can be conditioned in a number of ways.

2.4.2.1 Purge Gas Reforming

In principle, purge gas with methane contents up to 40 % can be reformed either endothermally or autothermally. A number of reasons, however, suggest that catalytic autothermal reforming will be the most appropriate route for purge gases from coal-based methanol synthesis plants; the reforming temperature range – both for catalytic autothermal reforming and for non-catalytic partial oxidation which takes places at a much higher temperature – is not limited by the reformer material, as it is the case with steam reforming. Catalytic autothermal reforming requires only the reforming catalyst and the ceramic bricklining material to withstand the reaction temperature, while for the non-catalytic partial oxidation process, only the bricklining has to be temperature-resistant. Hence, the catalytic autothermal reformer can be operated without problems at pressures around 50 bar and temperatures up to 1 000 °C.

The reforming temperature, which is some 100–150 °C higher for catalytic autothermal reforming as compared to endothermal reforming in a tubular reactor, leads to a more favourable CO/CO_2 ratio and a lower residual CH_4 content in

Table 2.6. Data on autothermic reforming of puregas

Purgegas from Methanol Synthesis:		
CO_2		5.0 vol.%
CO		3.0 vol.%
H_2		54.0 vol.%
CH_4		35.0 vol.%
N_2 + Ar		3.0 vol.%
Reforming Conditions:		
Steam/Purgegas Ratio	(mol/mol)	1.0
Pressure	(bar)	40.0
Temperature	(°C)	950.0
Oxygen / Purgegas	(mol/mol)	0.13
Reformed Gas:		
CO_2		4.33 vol.%
CO		13.96 vol.%
H_2		52.17 vol.%
CH_4		3.60 vol.%
N_2 + Ar		1.58 vol.%
H_2O		24.36 vol.%

the reformed gas. Raising the pressure requires nothing but a slight increase in reforming temperature. If for instance, the pressure for the example given in Table 2.6 is raised from 40 to 50 bar, an increase in the methane content of the reformed gas could be prevented by increasing the temperature from 950 to 975 °C. The oxygen demand for catalytic autothermal reforming is smaller than for non-catalytic partial oxidation which takes place at approximately 1 500 °C. The capital investment required for a catalytic autothermal reactor is only a fraction of the cost of a steam reformer; as most coal gasification plants require some unit to provide oxygen, the oxygen needed for autothermal reforming leads to only slightly higher cost for an air separation unit.

Catalytic autothermal reforming means that the purge gas from methanol synthesis is reformed autothermally, i.e. without external heat input, with oxygen and steam at pressures up to 50 bar and a temperature range between 950 and 1 000 °C. The reaction follows the laws of methane equilibrium, homogeneous water-gas equilibrium and *Boudouard* equilibrium as described already in Chapter 1. Since purge gas normally has a high hydrogen proportion and the *Boudouard* equilibrium is always established with some delay to the steam reforming reaction, autothermal reforming can do with a relatively low H_2O/C ratio. This leads to a lower oxygen demand and a more favourable CO/CO_2 ratio in the reformed gas for methanol synthesis without involving the risk of soot deposits in the reformer.

Figure 2.22 shows the simple configuration of a modern *autothermal reactor*. The purge gas is preheated to 400 to 500 °C and then enters the annular space, arranged concentrically around an inner tube, of the burner installed on

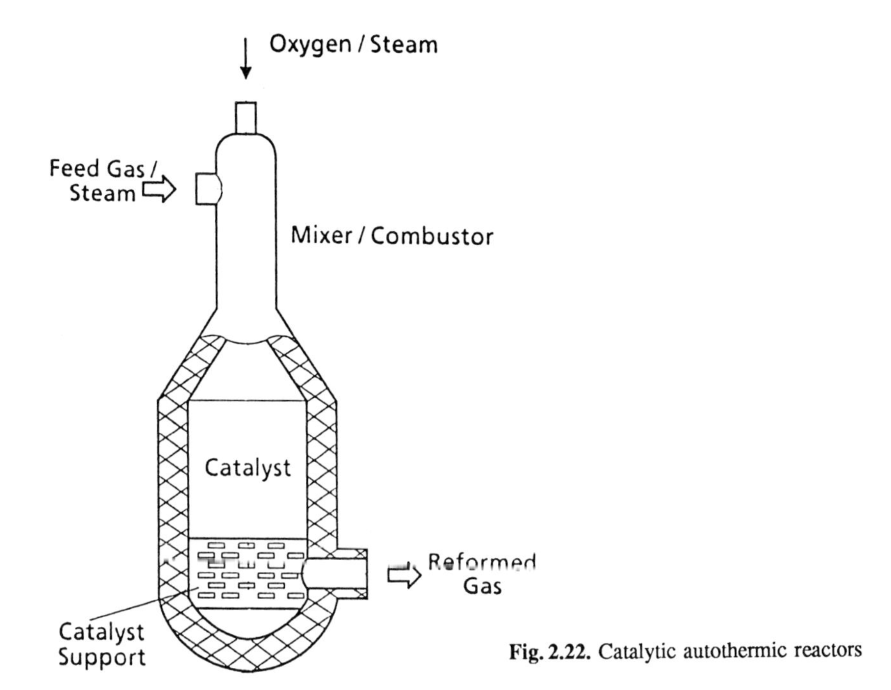

Fig. 2.22. Catalytic autothermic reactors

the reactor top. The oxygen/steam mixture which may also be preheated, enters through the inner tube and is thoroughly mixed with the purge gas leaving the burner, and combustion sets in already at this stage.

Temperatures as high as 1200 °C may be reached in this initial combustion zone above the catalyst bed. Steam reforming of methane also begins already in the initial combustion zone and, as this reaction consumes heat, leads to a considerable decrease of the gas temperature before it enters the catalyst. In the catalyst bed, the reaction then proceeds very fast, consuming heat, and the temperature in the catalyst bed drops swiftly to the reactor outlet temperature, which at this high level is identical with the methane equilibrium temperature.

The autothermal reactor is lined with ceramic materials. Two different materials are normally used: highly abrasion-resistant bricks on the inside, and materials with low-bulk weight and high-insulating capacity on the outside, i.e. facing the metal shell of the pressure vessel. A tight seal between the brickwork and the metal shell is of particular importance – metal rings welded to the pressure-containing shell at regular intervals are typically used to this end – because the pressure drop across the catalyst bed might otherwise give rise to parasitic slip streams between the brickwork and the pressure-containing shell, which may detract from the reforming efficiency and may even lead to soot deposits. The method practised in the past of enclosing such reactors by water-filled jackets is no longer used today, as modern insulating methods can safely prevent the dreaded hot spots. The catalyst-supporting grate is normally made of heat-resistant ceramic material as well.

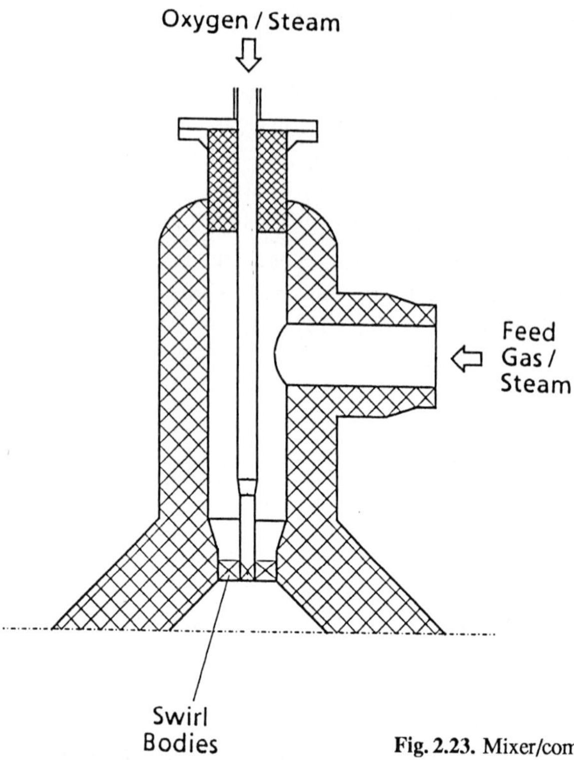

Fig. 2.23. Mixer/combustor of an autothermic reactor

The steam rate is appropriately set to ensure a difference of 100 to 120 °C between the reactor outlet temperature and the *Boudouard* equilibrium temperature. In addition to its moderating effect upon combustion in the initial combustion zone, the steam will then also safely prevent soot formation in the reaction zone and tends to reduce the residual methane content. The oxygen demand may be calculated from the system heat demand.

Figure 2.23 provides a schematic illustration of the autothermal burner or – to put it more correctly – mixer. Obviously, it is of the uncooled type which has in recent years, owing to the availability of high-temperature steels, prevailed over the water-cooled type. In order to ensure stability of the burner to the high temperatures in the initial combustion zone, it is important that the mixture of purge gas, oxygen and steam is ignited at a certain distance from the burner mouth, so that the metal parts are not exposed directly to the extremely high temperatures. This is ensured by sufficiently high flow rates of the reacting substances within the burner and by a custom-tailored design of the mixing elements in the burner mouth. The high flow rate within the burner also prevents backfiring into the purge gas supply line, but it also leads to a noticeable pressure drop from the burner inlet to the burner outlet. This pressure drop is normally recorded and its change is an early sign of irregularities in the burner unit.

The *catalysts* used for catalytic autothermal reforming contain some 10–14 % nickel. As a carrier material, pure aluminium oxide proved to be suitable also in high-temperature service. It is very important that the silicon-oxide content should be kept very low, as the substance becomes volatile in the presence of steam and may cause problems in the downstream equipment. 0.3 wt. % SiO_2 are normally considered the limit. The catalysts contain nickel in the form of nickel oxide which is reduced to metallic nickel when purge gas and gasifying agent are fed to the system after the catalyst has been heated to its operating temperature.

Table 2.6 lists the feed and reforming product from high methane purge gas, as well as the steam rate, oxygen demand and reaction conditions. A comparison between the H_2 quantity introduced by the purge gas and steam and the H_2 leaving the system in the form of reformed gas and residual steam shows that some 5.6 vol. % of the hydrogen in the purge gas are discharged in the form of water.

The cost-effectiveness of purge gas reforming by the catalytic autothermal process can be improved by using a saturator/desaturator system as described in Sect. 2.4. Such a system enables most of the process steam quantity required for the reforming reaction to be generated from the otherwise scarcely useable low-temperature heat. The heat available at a higher temperature level can then be used to generate high-pressure steam for export. If a medium-temperature gasification process is applied to certain type of coal, the H_2/CO ratio in the raw gas is only slightly less than desired for methanol synthesis. This means that in this case only very little CO needs to be steam-converted to H_2 and CO_2. Rather than shifting the raw coal gas, a hot clean gas shift conversion unit can then be used for the gas from a catalytic autothermal reforming unit. If the reformed gas of Table 2.6 were cooled to approximately 350 °C and then fed at this temperature to a single-stage high-temperature conversion unit, the steam content of the gas would permit the residual CO content to be adjusted to something like 5.0 vol. %. The cooled gas from the purge gas reforming loop would be fed to the CO_2 purification stage in the coal gas system. As complete recycling of the gas obtained from purge gas reforming would lead to an extremely fast build-up of nitrogen and argon in the methanol synthesis loop, a small purge gas bleed stream has to be withdrawn from the system and put to some other use.

Details concerning the integration of an autothermal reforming unit in a coal-based methanol plant will be described in Chap. 7, in which a general layout will be discussed.

2.4.2.2 Purge Gas Separation

If the purge gas from the methanol synthesis contains only little methane, as it is the case when high-temperature coal gasification is used, purge gas reforming will not be practical and it will be more appropriate instead to separate the valuable gas components from the undesirable ones either by pressure-swing absorption or by a prism-separator system before they are recycled to the main gas flow. The operating principles of the two systems have already been described in Chap. 2 above.

If a *PSA unit* is selected, the normal procedure is to separate a maximum of sufficiently pure hydrogen from the purge gas. All other gas components, including inert gases, are obtained together and have to be eliminated from the system. Small quantities of carbon oxides are also withdrawn together with the inert gases but this is normally not an economic handicap, as methanol-from-coal plants alway produce a great surplus of carbon. The drop in hydrogen pressure across the PSA Unit is slight, and the H_2 can be fed back into the synthesis loop at a suitable point without recompression.

The *prism-separator system*, too, is suitable to extract hydrogen alone with high purity from the purge gas. This system would also constitute a simple way of splitting a purge gas with high-methane content into a high hydrogen and a high-methane fraction. The former could then be recycled to the main gas system directly, while the latter would have to undergo catalytic autothermal reforming. Both gas flows obtained in this way have to be recompressed.

2.4.2.3 Partial Oxidation of Hydrocarbons and Oxygenates

Reforming the liquid hydrocarbons and phenols from coal gasification, like the reforming of high-methane purge gas, is a contribution to increasing the syngas yield from a given quantity of coal. A suitable means to this end is the SHELL Gasification Process, termed SGP for short or also POX (for *Partial Oxidation*), that was developed by SHELL *International Research Maatschappij*. This process was developed at the beginning of the fifties and is today used in numerous gasification plants for all types of hydrocarbons from natural gas to the heaviest oil residues such as propane asphalt [C25, C26]. Oxygenates of all kinds can also be partially oxidized together with hydrocarbons. The standard layout of an SGP plant is illustrated by Fig. 2.24.

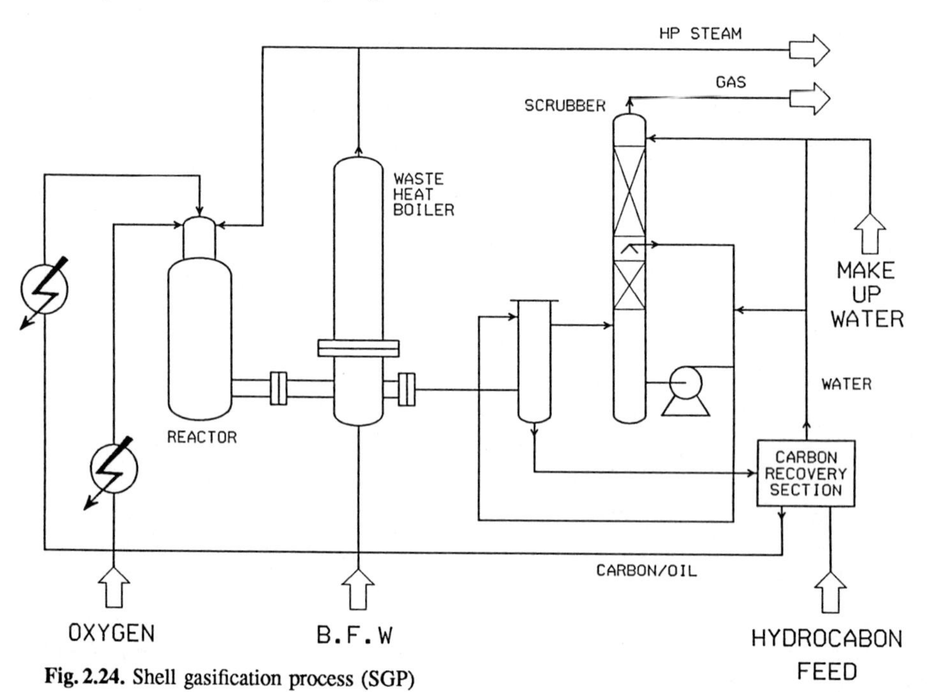

Fig. 2.24. Shell gasification process (SGP)

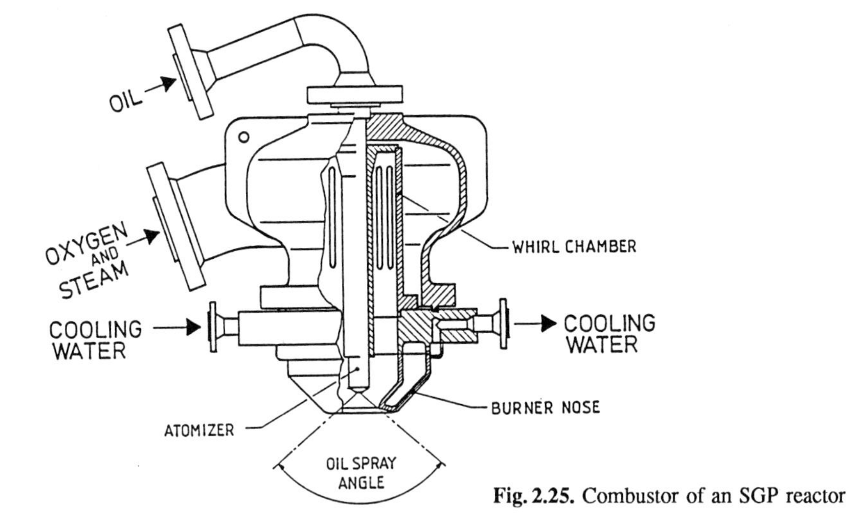

Fig. 2.25. Combustor of an SGP reactor

A reciprocating pump is used to raise the pressure of the liquid hydrocarbons to a level which, depending on the reactor size, is some 40–80 bar higher than the gasification pressure within the reactor[1]. The feedstock is then indirectly heated to a temperature at which its kinematic viscosity is approximately 20 cSt, similar to that of diesel oil at ambient temperature. Both the high pressure differential and the adjustment to a suitable viscosity are necessary to produce the very fine liquid particles when the feedstock is atomized into the combustor that will keep inevitable soot formation to a minimum. Figure 2.25 shows how the hydrocarbon gets into the reaction space through a gun at the centre of the combustor, while oxygen and steam are introduced through a swirl chamber and mixed with the hydrocarbon droplets at the burner mouth. Because of the high reaction temperature – the adiabatic flame temperature is as high as 1 400–1 500°C – the reactants are very rapidly converted into a gas with high H_2 and CO contents according to the reactions of partial and complete oxidation and to the water gas equilibrium mentioned already in Chap. 1 (Table 1.3). Methane formation follows the methane/steam equilibrium only in its tendency; in fact, the methane content of the reformed gas is about 10 times higher than expected for equilibrium. Figure 2.26 shows the theoretical methane contents of the reformed gas for the temperature and pressure range that is relevant for practical applications.

Although under the conditions prevailing in the reaction zone, neither the *Boudouard* equilibrium nor the heterogeneous water gas reaction would justify soot formation, free carbon is produced together with the reformed gas at a rate of 0.5–2 wt. % in terms of the feedstock, depending on the temperature and the quantity of steam added, but also on the retention time. When the soot is recycled together with the feedstock, the soot content in the reformed gas does

[1] Latest developments include a co-annular burner that allows differential pressures between burner inlet and reactor of 2–3 bar only.

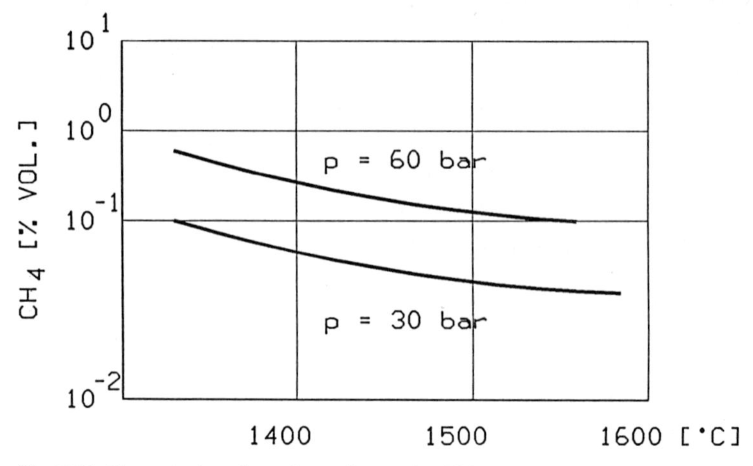

Fig. 2.26. Theoretical methane formation under SGP process conditions

not increase; this suggests that soot formation is not primarily due to thermal cracking of the hydrocarbons but to reaching the *Boudouard* limit when the gas is cooled down, the ash particles acting as condensation nuclei.

A specially designed waste heat boiler downstream of the reactor cools the gas to 300–350 °C, generating HP steam of as much as 120 bar. The special features of this boiler type are that the high gas flow rate in spiralling tubes has a self-cleaning effect and that a nicely controlled water flow around the tubes ensures that the wall temperature on the gas side can never rise to a point where the much-dreaded metal dusting, i.e. iron carbide formation, occurs which may cause devastating corrosion within an extremely short time. Under partial oxidation conditions, i.e. high CO partial pressure and low steam content in the reformed gas, this phenomenon will occur between 550 and 650 °C and it is caused by carbon radicals, which may be formed when the temperature drops below the *Boudouard* equilibrium, acting on the iron of the tube wall.

The gas leaving the waste heat boiler enters a quench cooler in which it is cooled below the dewpoint and most of the soot is eliminated. The condensing water particles act as nuclei for this soot removal. The nearly soot-free gas is then cooled to ambient temperature in a packed wash tower and the remaining soot is removed to a few ppm. The lower part of this wash tower operates on treated soot water which is thereafter sent to the quench cooler to wash out the major part of the soot. The top section of the tower is fed with fresh make-up water. The quench cooler effluent, containing 1–1.5 wt. % of soot, is discharged to a soot water treatment unit in which the soot – or, to be more accurate, the soot plus oil ashes – is removed from the water either by pelletizing it with feedstock oil or by agglomerating it by means of naphtha [2.27]. The low-soot water is recycled to the wash tower. The pelletized or agglomerated soot is normally remixed into the feedstock but may also be burnt under a boiler or in an incinerator. The latter method will be desirable if soot occurs only at a rate of 0.5 % in terms of the feedstock, or if the feedstock contains high percentages of ash – nickel, vanadium, or sodium.

If the partial oxidation unit is integrated in a coal gasification complex – as suggested in Chap. 7 – and the liquid products of coal gasification are used as a feedstock, the ash content of this feedstock will be extremely low and gasification can be performed in such a way that even less than 0.5 wt. % of soot occur in the reformed gas. In this case, the expenditure for a carbon recycling system can be saved as the small carbon quantities can be more easily and with less expenditure be removed in the waste water treatment system of the complex.

3. How to Synthesize Methanol and Alcohol Mixtures

3.1 Methanol – Its History, Its Properties and What Becomes of It

3.1.1 The History of Methanol

Methanol has a history extending back to the year 1661. *P. Boyle* then succeeded for the first time in recovering methanol from the components of wood vinegar. He termed the product obtained by rectification "*adiaphorus spiritus lignorium*". As at that time there was no use whatsoever for the substance methanol, it was forgotten for many years until *Taylor* rediscovered it in 1822. *Justus von Liebig* succeeded in clarifying the chemical individuality of methanol in 1835. He gave it, for the first time, the correct chemical formula

CH_3OH

and introduced it into chemistry as methyl alcohol.

In the hundred years following this, methanol was recovered to an increasing degree as *wood alcohol* by dry distilling wood. After the synthetic production of methanol from carbon oxide and hydrogen was introduced, wood distillation was very quickly displaced and today it amounts to less than 0.1 % of the world methanol production.

A. Mittasch and his staff succeeded in first producing methanol from carbon monoxide and hydrogen in 1913. Iron was used as catalyst. The disadvantage of this process method was the fact that the methanol was recovered together with a whole series of other compounds containing oxygen and that the catalysts used for synthesis only had very short cycle times. In 1921, *M. Patart* described a methanol synthesis process in which hydrogenation active metals and metal oxides were stated to be the catalyst. The first commercial methanol plant, even if on small scale, based on his process was built in France in 1922.

The rapid rise in the production of methanol as chemical raw material first began in 1923 when the first large-scale synthesis plant was constructed by the *Badische Anilin und Sodafabrik*, Ludwigshafen (BASF). Based on the work by *Mittasch*, *M. Pier* and his staff proved during test covering several years that iron is scarcely suitable as a catalyst for methanol production. On the contrary, it was established that iron must be carefully avoided in the production of methanol synthesis catalysts as its presence under the operating conditions applied to methanol synthesis leads to the formation of iron penta carbonyl which on its part resulted in iron deposits on the catalysts and methane being formed from the synthesis gas. *Pier* developed *chrom /zinc oxide* catalysts which were suited for methanol synthesis and also had the advantage that their sensitivity to sulfur is low. The first large-scale commercial plant using the process developed by *Pier* operated at a pressure of 350 bar and at temperatures around 400 °C. A series of commercial plants went into operation at the beginning of the thirties in the USA. The capacity of these plants was approximately 100 to 500 tons per day and chromic acid activated zinc oxide catalysts were used almost exclusively.

As early as 1935 it was recognized that the application of catalysts on a copper basis provide considerable advantages for synthesizing methanol as against the zinc oxide/chrome oxide catalysts. The first patents covering the manufacture and application of *copper catalysts* for methanol synthesis were applied for in 1937 and the best known of these containing all the elements of the copper catalysts customary today was Polish and submitted by *Pospekhov*. Catalysts on a copper basis permit methanol synthesis at considerably lower pressure and, above all, lower temperatures. The disadvantage of the copper catalysts is that they are extremely sensitive to sulfur. Only from the end of the fifties, the development of suitable processes for purifying either the raw materials intended for methanol synthesis or synthesis gases containing sulfur made it possible to apply copper catalysts for methanol synthesis.

Tests with catalysts containing copper were carried out by *Imperial Chemical Industries Ltd.*, England, from about 1958 to 1962 and eventually a practical copper catalyst for methanol synthesis and the first *Low-pressure Methanol Processes* were brought onto the market. In the process developed by ICI the quench reactor, in which the reaction heat is removed by quenching with cold gases and which is known from high-pressure methanol synthesis, is used.

Methanol synthesis at low pressures and temperatures presents many advantages as against the old high-pressure processes. The most important of these advantages include the considerable saving in energy and the much lower formation of byproducts. At about the same time as ICI announced the first low-pressure methanol plant in the world was going into operation at Billingham, Messrs. *LURGI Gesellschaft für Wärme- und Chemotechnik* was also developing a low-pressure methanol process. Contrary to the quench reactor used by ICI, LURGI applied a *tubular reactor* cooled with boiling water, which made the energy advantages of the low-pressure process particularly noticeable. Most of the methanol plants built in the last 20 years operate to the ICI or LURGI processes and numerous high-pressure synthesis units have been converted to low-pressure methanol processes in the second half of the last decade.

With the increasing diversification in the application of methanol, primarily in view of using methanol as motor or other fuels, it may well be expected that plants will be built very soon with capacities far above today's maximum size of about 2 500 tpd.

This brief history of methanol production would not be complete without two contributions to be found in reference works published at the turn of the century. It is mentioned that the first application for methanol, then still termed *wood alcohol*, was the denaturing of spiritus. For this purpose, wood alcohol was highly appropriate due to its particularly intensive odour. At that time, methanol was also used for manufacturing paints and polishes as well as formaldehyde and methylether (used as refrigeration agent). In addition it is to be read that in 1980 the price for methanol was about 50 times as high as it is today if one compares it on the basis of the same price level.

3.1.2 The Use of Methanol

Numerous studies and publications exist about methanol producing facilities world-wide, indicating their on-stream rates, and on the methanol market of past and coming years [3.1,2]. It should only be mentioned here, that methanol, having an installed production capacity of approximately 15 million tons per year (1984) is one of the most important chemical raw materials.

Whereas most of the methanol produced since the thirties until 1980 has been used to produce formaldehyde, a remarkable shift in the pattern of methanol use has occurred. Nowadays, use of methanol for chemical products other than formaldehyde has risen more steeply than for formaldehyde itself. More than all others, the increase of acetic acid production going together with its shifting from ethylene to methanol and carbon monoxide as raw materials has contributed to this increase as well as the production of fuel components such as MtBE.

In 1982, the following typical use of methanol was published for the USA consumption amounting to 3.1 million tons:

Formaldehyde	30.5 %
MtBE and other fuel uses	13.3 %
Acetic acid	11.6 %
Solvents	10.5 %
Chloromethanes	8.7 %
Others	25.4 %

It has been predicted that until 1990 the methanol consumption will increase by 30 % as compared to 1984. This would mean 19.5 million tons per year to be produced in 1990, largely influenced by the increasing use of methanol in the fuel sector.

3.1.3 Physical Properties

Methanol, or as it is termed in full *methyl alcohol*, with the chemical formula CH_3OH is the first of the long series of alcohols. Its molecular weight is 32.04 and it is a neutral, colourless liquid in pure condition having an odour similar to that of ethyl alcohol. Methanol dissolves well with other alcohols, esters, ketones as well as with aromatic hydrocarbons and water. It can be less well mixed with fats and oils. It dissolves a number of organic substances including numerous salts. The most important physical data for methanol are assembled in Tables 3.1 to 3.5. Further data on the physical properties of methanol can be taken from the literature under [3.3–to 3.10].

Table 3.1. Physical data of methanol

Boiling Point (1013 mbar)	64.509	°C
Melting Point	-97.88	°C
Gravity d_4^{0}	0.81009	
d_4^{25}	0.78687	
d_4^{45}	0.76761	
Specific Heat (liq.) at -125.30°C	1.6953	J/g K
- 51.47°C	2.2325	J/g K
- 6.15°C	2.3698	J/g K
0.42°C	2.3920	J/g K
19.85 C	2.4979	J/g K
Heat of Combustion at 25.0 °C	22.6926	kJ/g
Heat of Evaporation at 20.0 °C	38.48	kJ/mol
at 64.7 °C	35.346	kJ/mol
Melting Heat	99.23	J/g
Free Energy of Formation		
$F_{298.16\ K}$	-162.03	kJ
Entropy (liq.)	126.9	J/mol K
Entropy (gas) at 298.16 K	241.46	J/mol K
Critical Data		
Temperature	239.43	°C
Pressure	80.92	bar
Gravity	0.272	g/cm^3
Flame Point acc. ABEL-PENSKY	6.5	°C
Ignition Temperature	470	°C
Explosion Range in Air	6.72 - 36.5	% vol.

Table 3.2. Vapor pressure of methanol

°C	mbar
- 16	13.3
- 6	26.9
0	39.7
10	73.2
20	128.6
30	216.7
40	352.0
50	552.0
60	841.0
70	1246.0
80	1801.0
90	2546.0
100	(3500.0)

Table 3.3. Specific gravity of methanol-water mixtures at 25°C

Methanol % weight	Methanol % vol.	Spec. Gravity d_4^{25}
0	0.00	0.9971
10	12.46	0.9804
20	24.53	0.9649
30	36.20	0.9492
40	47.37	0.9316
50	57.98	0.9122
60	67.96	0.8910
70	77.19	0.8675
80	85.66	0.8424
90	93.33	0.8158
100	100.00	0.7867

Table 3.4. Specific heat of methanol vapor at 1013 mbar

°C	J/mol K
74.2	86.92
83.4	66.66
100.2	56.23
125.8	53.01
128.0	54.05
158.3	55.81
169.0	56.06
184.2	57.30
204.6	57.28
211.9	56.61
225.8	60.17
248.2	61.59
282.8	63.98
308.2	66.40
312.2	66.86

Table 3.5. Freezing point of methanol-water mixtures

Methanol (% by weight)	Freezing Point (°C)
0	0
10	- 6.5
20	-15.0
30	-26.0
40	-39.7
50	-55.4
60	-75.7
64	-83.4

3.1.4 Chemical Properties

Methanol decomposes to formaldehyde when it is subject to the ultraviolet radiation contained in the sun light, even extraordinarily slowly. The direct oxidation on suitable catalysts and at temperatures of 650 to 1 000 °C yields formaldehyde. Dimethyl ether is produced from methanol by dehydration.

Methanol can be esterified using a mixture of sulfuric and nitric acid, converted to methyl chloride with hydrogen chloride and to methyl amines with ammonia. Alkali metals are dissolved by methanol with alkali methylates forming; if this reaction takes place in the presence of CO, methyl formiate occurs. One of the most important reaction occurring in the presence of carbonyl catalysts is CO being taken up by methanol producing acetic acid.

3.1.5 Toxicity

Methanol can damage the human organism. Poisoning can occur due to inhaling vapours and by orally taking liquid as well as due to long contact with the skin. Slight poisoning is first made apparent by intoxication-like symptoms, then by headache and nausea, often accompanied by visual disturbance, bad cases can led to blindness and occasionally be deadly due to paralyzing the respiratory track. Methanol is only broken down by the body slowly, accumulative effect thus being able to occur on small quantities being constantly taken. The odour threshold in the air breathed for methanol is about 0.3 to 0.5 vol. %, the maximum concentration allowed at a place at work is stipulated at 200 ppm, i.e. 0.02 vol. %.

The individual sensitivity in regard to oral intake of methanol appears to be just as different as in case of the intake of vapours. Cases are known where the intake of even few mg led to death, whereas quantities of 250 mg have been survived, however, the worst of toxication symptoms being exhibited. Poisoning is caused in that in the body methanol further oxidizes initially to formaldehyde and then to formic acid causing overacidification of the blood.The administration of ethanol is practiced as first aid inhibiting the biological methanol oxidation and thus suppressing the formation of formaldehyde or formic acid.

3.2 Fundamentals of Methanol Synthesis

3.2.1 Physical Basis

Today methanol is almost exclusively produced by catalyzing carbon monoxide and carbon dioxide with hydrogen. Formation is to the mechanism

$$CO + 2H_2 \rightleftarrows CH_3OH$$

and

$$CO_2 + 3H_2 \rightleftarrows CH_3OH + H_2O$$

heat occurring during the conversion.

The change in the basic enthalpy is in respect of the following formulae

$$H_{25} = 90.84\,\mathrm{kJ/mol} \quad \text{and}$$

$$H_{25} = 49.57\,\mathrm{kJ/mol}$$

respectively.

The fact that both reactions are exotherm and in addition, proceed under volume contraction shows that the highest conversions – and thus the highest methanol yield – are favoured by low temperatures and high pressures.

The conversions are limited by the position of the chemical equilibrium. The temperature dependency of the equilibrium constant K_1 for converting CO and H_2 can be expressed by the following equation [3.11]:

$$\log k_1 = 3921/T - 7.971 \log T + 2.499 \cdot 10^{-3} \cdot T$$
$$- 2.953 \cdot 10^{-7} \cdot T^2 + 10.20 \quad .$$

The second reaction is coupled to the first one by the water gas equilibrium

$$CO + H_2O \rightleftarrows CO_2 + H_2$$

It can be most practically determined as K_2/K_3, K_3 representing the equilibrium constant which can, for example, be described as follows [3.12]:

$$\log k_3 = 2203.24/T - 5159 \cdot 10^{-5} \cdot T - 2543 \cdot 10^{-7} \cdot T^2$$
$$+ 7.461 \cdot 10^{-11} \cdot T^3 + 2.3 \quad .$$

CO conversion is preferred to the CO_2 conversion by all copper catalysts applied in practice. Therefore one endeavours to produce methanol synthesis gas with the highest possible CO content and lowest possible CO_2 content.

This is possible in case of all synthesis gases produced from coal up an H_2/CO ratio equivalent to the theoretically optimal stoichiometric number of

$$SN = (H_2 - CO_2)/(CO + CO_2) = 2.0$$

in the absence of CO_2.

However, in practice a certain minimum quantity of CO_2 must be present in the synthesis gas to attain a high CO conversion. Generally, a maximum conversion is attained – at least this applies to all catalysts examined by the author – when the CO_2 content in the synthesis gas is between 2.5 and 3.5 vol. %. To data, this phenomenon could not be convincingly explained. Whereas some specialists are of the opinion that methanol formation is only effected by the indirect route of the CO_2/H_2 conversion, others are of the opinion that CO_2 in the synthesis gas represents a type of initial trigger for the CO conversion. Others again believe that CO_2 in small quantities acts on the catalyst as promotor, so to speak. On the CO_2 content in the synthesis gas rising, the CO conversion goes up to a maximum and then in turn drops initially flatly and then steeply. This behaviour is shown in Fig. 3.1 as a typical case. In regard to several commercial

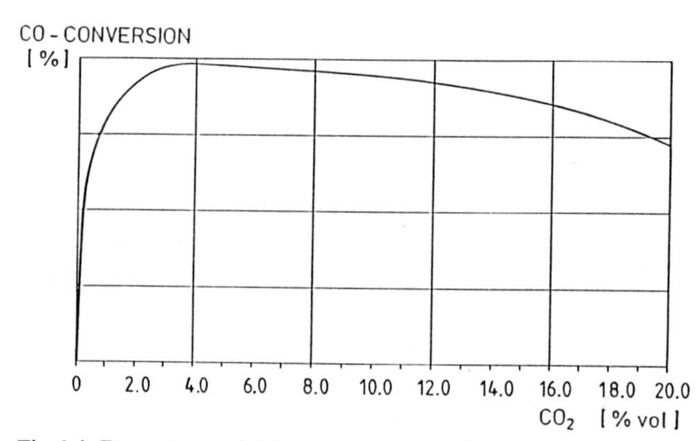

Fig. 3.1. Dependence of CO conversion upon CO_2 concentration in reactor feed gas

Table 3.6. Conversion of CO_2 and CO at equilibrium*

Pressure	25 bar		50 bar		70 bar	
Temperature	$CO_{conv.}$	$CO_{2conv.}$	$CO_{conv.}$	$CO_{2conv.}$	$CO_{conv.}$	$CO_{2conv.}$
275°C	0.130	0.043	0.363	0.051	0.506	0.059
300°C	0.052	0.060	0.196	0.070	0.316	0.079
327°C			0.084	0.097	0.156	0.107
337°C					0.114	0.120

* Starting from Gas containing 3 % vol CO_2, 27 % vol CO, 64 % vol H_2, 6 % vol CH_4+N_2

copper-based catalysts, it could be proved that no methanol can be produced using synthesis gas containing no CO_2 and from which all water was withdrawn to prevent CO converting to CO_2.

The conversion of CO and CO_2 to methanol is limited by the pertinent chemical equilibria. The conversions of CO and CO_2 resulting at various pressures and temperatures on the equlibrium being adjusted were listed in Table 3.6 for a synthesis gas produced from coal containing 3.0 % CO_2, 27.0 % CO, 64 % H_2, 4 % CH_4 and 2 % N_2, all by volume.

An unlimited residence time of the reaction partners under reaction conditions would be necessary to adjust the methanol equilibrium, CO_2 or CO being the basis. Transferring this to the practical application case it means that one would require an unlimitedly large reaction chamber or such a catalyst volume. Therefore on determining of the catalyst volume one tries to approach the methanol equilibrium from the CO conversion, which – as already mentioned – is kinetically favoured, only so far that further conversion increases would be obtained with large catalyst increments only.

Figure 3.2 shows the approach of the CO conversion to the equilibrium at a highly active copper catalyst applying the residence time or – in other words –

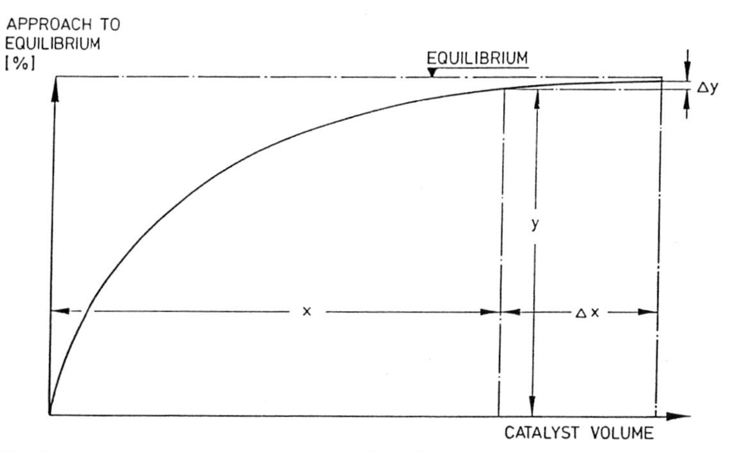

Fig. 3.2. Excessive catalyst requirement (Gdx) for marginal improvement of approach to equilibrium

the catalyst volume. One recognizes that on having reached point Y at a catalyst quantity X, the catalyst volume would have to be increased by a δX of 35 % in order to obtain an increase of Y by a δY of only 2 %. The approach to the equilibrium for CO conversion on modern methanol catalysts is above 99 % and for CO_2 conversion about 94 %.

3.2.2 The Methanol Yield from Synthesis Gas

In the above section, the importance of carbon monoxide and carbon dioxide conversion and the technically attainable approach to the equilibrium has been described. However, these two parameters alone do not decide upon the optimation for the production of methanol from a specific synthesis gas. The methanol yield from the synthesis gas is of quite decisive importance for economically producing methanol on a commercial scale. Its this yield on which depend the quantity of synthesis gas which must be produced from coal, cleaned, conditioned and compressed and the quantity of CO_2, CO and H_2 which must be removed from the methanol synthesis as *purge gas* and thus is lost to methanol production by the direct route.

So far talk was always of synthesis gas and it was implied – in silence – that this gas enters a reaction vessel filled with catalyst in which CO_2, CO and H_2 are then converted to methanol and that during this a gas mixture occurs containing methanol, residual CO_2, residual CO and residual H_2 approximately in equilibrium as well as all the inert gases contained in the synthesis gas. If one were to apply the coal gas described in the previous section containing by volume 3.0 % CO_2, 27 % CO and 64 % H_2 at a pressure of 70 bar and a temperature of 250°C in a reaction vessel filled with a methanol catalyst, the conversion of CO_2 and CO would commence vehemental and be accompanied by just as vehement a temperature rise. Assumed that methanol equilibrium was reached, a gas mixture would be available at the end of the reaction at a temperature of 337°C and with a composition as follows:

CH_3OH	3.70 vol. %
CO_2	2.84 vol. %
CO	25.68 vol. %
H_2	60.94 vol. %
CH_4	4.30 vol. %
N_2	2.15 vol. %
H_2O	0.39 vol. %

Only 12.0 vol. % of the CO_2 applied and 11.4 % of the CO introduced would have been converted to methanol and 1 volume synthesis gas would have shrunk to 0.93 remaining gas volume due to the volume concentration which took place. Thus 28.74 mol synthesis gas would be required to produce 1 mol methanol or – to speak on commercial terms – to manufacture 1 ton of methanol, 20 000 m^3 synthesis gas would have to be provided.

This example shows that for commercial methanol production, two measures are imperative:

- The temperature rise inherent to the reaction must be minimized in order to operate at good equilibrium values. Later it will be shown that this is also of importance for retaining the activity of the copper catalysts applied these days.
- Carbon dioxide and carbon monoxide must be converted as far as possible, i.e. consumed, to attain the highest possible methanol yield from synthesis gas.

Both are attained if one dilutes the synthesis gas before it enters the reaction chamber. The most practical agent for this dilution is the reacted gas leaving the reaction chamber after the methanol formed was separated from it.

It appears appropriate to more clearly define the term synthesis gas at this point and introduce two new terms. *Synthesis gas* is the gas coming from synthesis gas production or cleaning which is fed to the methanol synthesis. The gas entering the methanol reactor should in future be termed *reactor feed gas* and that leaving the reactor *reacted gas*. When one has separated the methanol formed (and water) from the reacted gas, it is called *recycle gas* and the residual gas stream leaving the methanol synthesis and containing the inert gases contained in the synthesis gas as well as non-converted percentages of CO_2, CO and H_2 is termed *purge gas*. In order to maintain the temperature rise within acceptable limits, a dilution ratio of more than 30 : 1 would have to be selected, i.e. some of the synthesis gas would have to be mixed with 30 parts recycle gas. However, this would reduce the CO_2 and above all the CO content so far that the active concentration gradient of the main component CO largely becomes lost.

As this active concentration gradient is primarily decisive for dimensioning the necessary catalyst volume, one is satisfied with a compromise in practice and selects a dilution ratio of 3 : 1 to 7 : 1. Impermissible temperature rises on permanently arranged catalysts are controlled in that

- one divides the catalyst volume into several sections and adjusts the desired inlet temperature for the next section in each case by quenching with cold gas down-stream of each section, or
- one uses the reaction heat in a heat exchanger arranged in the catalyst chamber for preheating the reactor feed gas, or
- the reaction heat being released is utilized applying a cooling system accommodated in the catalyst chamber, e.g. for evaporating water.

In case of processes operating with catalyst systems where pulverized catalyst is circulated with the aid of a carrier liquid (*liquid entrained system*) or small-lump catalyst maintained in suspension in a liquid *(liquid fluidized system)*, it is possible to transfer the reaction heat inside the reactor to the carrier liquid and to remove it outside of the reactor. The various means of temperature control inside and outside the catalyst chamber are schematically shown in Fig. 3.3.

If one now goes back to the figure play dealt with in this section but assumes that due to suitable heat removal the temperature rises during the reaction procedure only from 250 to 255°C and that per volume synthesis gas four vol-

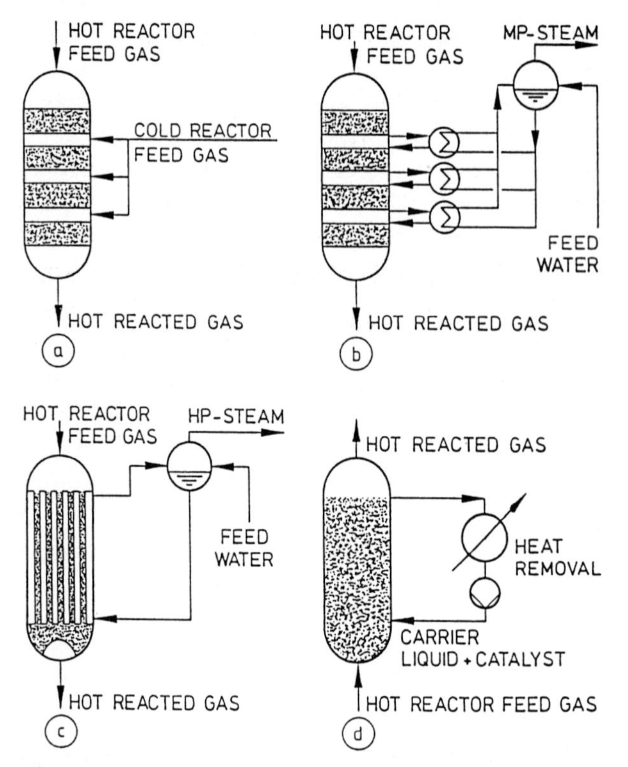

Fig. 3.3a–d. Various types of methanol synthesis reactors. **(a)** Cold gas quench; **(b)** cooling by evaporation – multistage, adiabatic; **(c)** cooling by evaporation – tubular, near isothermal; **(d)** liquid entrained system using heat carrier liquid

umes recycle gas are mixed, one obtains – equilibrium adjustment also being assumed here – 4.45 volumes reacted gas with a methanol content of 6.7 vol. %. Now only 2 560 m_3 synthesis gas is required per ton of methanol and the purge gas only contains 5.1 % of the CO contained in the synthesis gas. In the case of our once through example it was 88.6 %.

3.2.3 The Catalysts for Methanol Synthesis

Whereas formerly *zinc oxide/chrome oxide* catalysts were in general applied for methanol synthesis under high pressures – about 300 to 400 bar – which featured high temperature resistance and relative insensitivity to catalyst poisons, above all sulfur in many forms, today catalysts on a *copper basis* are used exclusively. Sometimes they are termed *Blasiak catalysts* in specialised literature and a number of works [3.14] exist on the first test results with this catalyst type. The copper-based catalysts permit methanol to be synthesized in an economic manner at pressures between 50 and 100 bar and temperatures around 230 to 270°C. The fact that it was first applied commercially about 1970 was mainly due to people not being in a position in the forties and fifties to obtain the necessary gas purity – above all sulfur contents less than 0.1 ppm vol. – using economic gas cleaning processes. In the early days, however, excessively high test temperatures were also frequently selected, indeed with the purpose of improving the reaction kinetic, and thus caused rapid recrystallization of the copper leading to disactivation. The recrystallization of the copper starts theoretically even at the *Tamman temperature* which is about 190°C [3.15]. The movements in the crystal lattice do then, however, still progress very slowly and not until temperatures round 270°C does recrystallization progress so quickly that it is no longer possible to apply copper catalysts economically above this temperature level.

The first copper catalysts suitable for methanol synthesis on a commercial scale were developed at the end of the 60s by ICI in Billingham [3.14] and LURGI in Frankfurt [3.15]. According to the first relevant patents owned by the two companies, these catalysts had the following composition:

Atom %	ICI	LURGI
Cu	90–25	80–30
Zn	8–60	10–50
Cr	2–30	–
V	–	1–25
Mn	–	10–50

During the course of constant catalyst development, attempts were made by adding promotors to improve the activity and thermostability and increase the selectivity, i.e. largely to suppress byproduct formation. During these efforts it was recognized that the method of manufacturing the catalysts also has decisive influence on their characteristics and that those components containing these impurities in the metals used for catalyst production are of great importance too, above all in regard to byproduct formation. Thus, the type of precipitation has considerable influence on the pore structure and size, for example.

The question of selectivity is therefore of great importance as forming methanol from CO, CO_2 and H_2 is little favoured thermodynamically. The following correlations between metallic catalyst impurities or improperly applied promotors and the formation of undesirable substances accompanying the methanol are known:

- Alkali and alkaline earth oxides promote the formation of higher alcohols, above all ethanol.
- Iron, cobalt and nickel reinforce the formation of higher paraffinic hydrocarbons; even a slight methanization proceeding along with the methanol synthesis was observed on some catalysts.
- Aluminium oxide as promotor can increase dimethyl ether formation considerably, above all at the top end of the temperature range approaching the copper catalyst application limit.

The following catalyst systems are known from patents issued from 1968 to about 1975 [3.8]:

ICI	Cu, Zn, Cr, Cu, Zn Al
LURGI	Cu, Zn, Cr, Cu, Zn, Mn, V
CCI	Cu, Zn, Al
BASF	Cu, Zn, Mn, Cr, Cu, Zn, Mn, Al, Cr, Cu, Zn, Al
SHELL	Cu, Zn, Ag
Mitsubishi	Cu, Zn, Cr

The catalysts supplied commercially these days are usually manufactured in cylindrical or tablet form measuring between 4 and 7 mm and the bulk weight is between 1.0 and 1.5 kg/l. The catalysts are applied in the pressure range from 50 to 100 bar and at temperatures between 230 and 280°C. At space velocities up to 12 000 m^3/h, time space yields of 0.5 to more than 1.2 kg of methanol/liter of catalyst are attained.

The conversion behaviour of the catalysts can only be determined empirically. During a number of tests with varying

- composition of the reactor feed gas
- reactor pressure
- space velocity (residence time)

it must be determined in how far a catalyst is in a position under differing conditions to reach the methanol equilibrium in each case and at what speed these reactions progress.

An equation form describing the behaviour of catalysts in general [3.16] and methanol catalysts in particular [3.17,18] and determination of reaction speeds [3.19] appears regularly in publications. It can be formulated in general as follows:

$$r = \frac{f_{CO} \cdot f_{H_2}^2 - f_{CH_3OH} \cdot p_{CH_3OH}/k_{eq}}{(A + B \cdot f_{CO} \cdot p_{CO} + C \cdot f_{H_2} \cdot p_{H_2} + D \cdot f_{CH_3OH} \cdot p_{CH_3OH} + E \cdot f_{CO_2} \cdot p_{CO_2})^3}.$$

Herein are r = reaction rate (kmol/h · kg of catalyst); f = fugacity coefficient; p = partial pressure (atm); A, B, C, D, E = kinetic constants; k_{eq} = equilibirum constant. The constants A, B, C, D and E must be determined for each catalyst.

The copper contained in the catalyst is in oxidic form, not catalytic active, due to the method of manufacture. Therefore, these catalysts must be reduced before beginning synthesis operation and thus the copper converted to its metallic, catalytic active form. This is done in that reduction proceeds highly exothermally with the aid of a circulated inert gas dosed with small quantities of hydrogen until no further hydrogen consumption can be determined. Reduction is carried out at low pressure and the temperature is gradually increased from about 150°C at the beginning to about 250°C.

As copper catalysts tend very greatly to oxidation in reduced condition at elevated temperature and the overheating occurring during emptying can be a source of danger to equipment, operating personnel or the vicinity, they are oxidized and cooled in situ analogously to the reduction procedure described above but with air carefully dosed to the inert gas being circulated or by passing nitrogen with a small volume percent of oxygen through.

3.2.4 Selection of Reaction Conditions

The reaction conditions for methanol synthesis must be optimally matched to keep the costs for the equipment and – even more important today – the raw material demand low. The process variants decisive for such optimization are

- composition of the synthesis gas (stoichiometric number, H_2, CO and CO_2 content, inert gases)
- synthesis pressure (as well as the pressure ratio between synthesis and gas generation)
- reaction temperature
- space velocity (catalyst load)
- recirculation ratio (recylce gas/synthesis gas).

The positive and negative influences of the process variations must be weighed up against one another to reach the aim of the optimization within the synthesis, namely obtaining as much methanol as possible from a specified quantity of synthesis gas at the lowest consumption of energy and the lowest possible plant cost.

The requirements for the optimum synthesis gas can be formulated quite clearly. It should have a stoichiometric number of 2.02 to 2.04 at a CO_2 content of about 3 vol. %, i.e. as much CO as possible, and it should contain little inert gas. As a general rule, one leaves a certain excess of hydrogen in the synthesis gas as this leads to an improvement in the space time yield. This is caused by the facility of copper catalysts to more intensively adsorb carbon oxides on their surface than hydrogen and therefore a limited surplus of the latter is an advantage. In addition, this can counteract excessive concentration of inert gases in the loop and an excessive increase in the CO content of the circulating gas also. An excessive inert gas content would mean reaction inhibition and on the CO content rising a clear increase in byproduct formation, above all in regard to higher alcohols, will be observed.

The majority of coal gasification processes permits the requirement regarding CO_2 percentage and stoichiometric number to be easily fulfilled, even the desired hydrogen surplus can easily be set. However, higher contents of inerts, in general methane, must be accepted in the case of certain gas production processes except when the raw gas from the coal gasification was postreformed with oxygen. Under certain circumstances even this can prove economic if, for example, there is no application for large quantities of purge gas with a high methane content, as is found downstream of the synthesis without oxygen postreformation or the size of the gas production and gas cleaning units can be considerably reduced by this. If, for example, a raw gas from a LURGI coal pressure gasification unit containing 10 vol. % CH_4, 3 vol. % CO_2 and having a stoichiometric number of 2.04 after cleaning and conditioning were to be supplied to the methanol synthesis unit, one would require 2765 m^3 synthesis gas per ton of methanol, and about 586 m^3 purge gas with about 44 vol. % CH_4 are obtained. If one submits the same raw gas to catalytic reforming with oxygen and during this reduces the

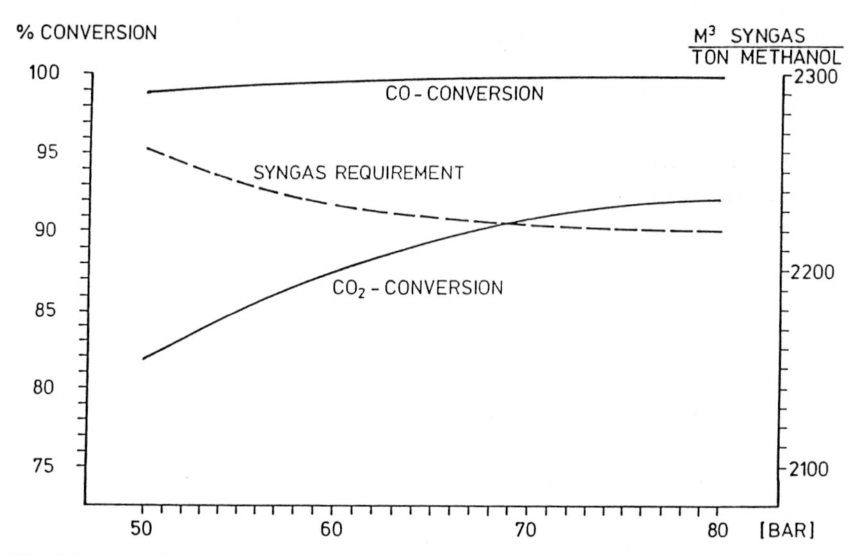

Fig. 3.4. Conversion of carbon oxides and syngas requirement depending upon pressure

CH_4 content to 0.5 vol. %, the synthesis gas demand can be reduced to 2 260 m^3 and the purge gas incidence to 85 m^3 per ton of methanol under the same process conditions in the synthesis loop. Gas production, gas cleaning and synthesis loop would thus be about 18 % smaller, the oxygen requirement in the gas production section would, however, practically double.

Rising pressure favours CO and CO_2 conversion. However, even at about 60 bar, more than 99 % of the carbon monoxide in the synthesis gas is converted on good catalysts and only the conversion of carbon dioxide still increses notably at higher pressure. The CO and CO_2 conversion rates attainable betwen 50 and 80 bar for a synthesis gas from coal with optimum composition are shown in Fig. 3.4 as is the requirement for synthesis gas per ton methanol. Figure 3.5 illustrates the rising energy requirement for synthesis gas and circulating gas compressors; the figures are based on the assumption that the synthesis gas is available at 25 bar downstream of the gas cleaning stage. As is to be seen from the two figures, the synthesis gas requirement does, indeed, drop by about 1.8 % if the pressure in the synthesis stage is increased from 50 to 80 bar but at the same time, the energy requirement for the gas compression rises by 22 %.

The costs for the equipment in the synthesis loop and the synthesis gas compressor move in opposite directions on the pressure in the synthesis rising. Whereas equipment, heat exchangers and piping can be sized smaller on the pressure rising, the costs for the synthesis gas compressor and, above all, often those for the pertinent drive turbine rise. Modern coal gasification processes operating at pressures around 60 bar and above permit synthesis gas compression to be dropped fully without the economics of methanol production suffering. A plant without synthesis gas compression and a capacity of about 240 000 t/y has been in operation in the Federal Republic of Germany since 1972. The synthesis

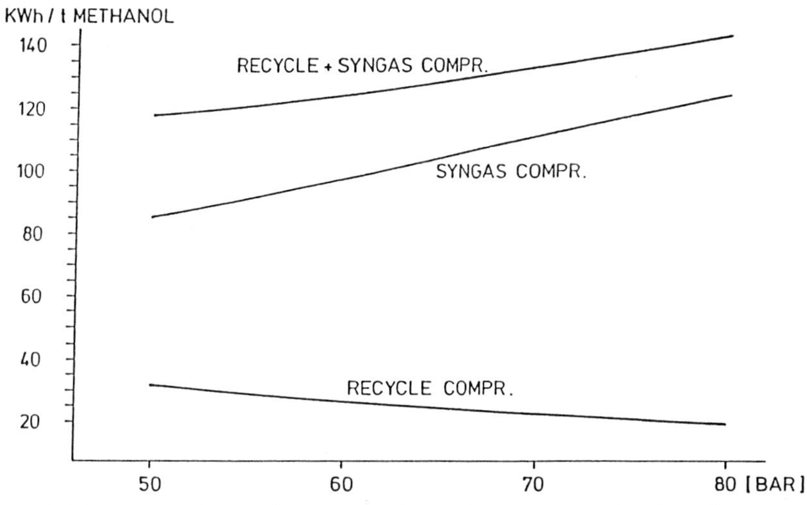

Fig. 3.5. Energy requirement for syngas and recycle gas compression depending upon synthesis pressure

gas, however, came from partial oxidation of vacuum residue oil exhibiting the same characteristics as does optimally conditioned gas produced from coal.

As the copper catalyst is only slightly active at temperatures below 230°C and accelerated recrystallization of the copper must be expected at temperatures exceeding 270°C, there is only little tolerance in selecting the most favourable reaction temperature. Low temperatures favour the equilibrium of methanol formation and elevated temperatures cause an improvement in the reaction kinetics. Both factors are utilized in that the methanol catalyst is operated at the lowest possible temperature as long as it still exhibits its full activity and only on this activity dropping is the reaction temperature increased. The deterioration in the methanol equilibrium at the end of the reaction caused by the temperature increase can be counteracted by increasing the recycle ratio, i.e. greater dilution of the reaction gas. The endeavour of optimally uniting high reaction speed and favourable methanol equilibrium is most closely accommodated by the quasi-isothermal reactor type. As is to be seen from the temperature profile shown in Fig. 3.6, it differs from the reactors with direct or indirect intercooling in that high temperatures occur in those place where high reaction speed is desirable, namely in the top section of the reaction chamber, and low temperatures establish where the methanol equilibrium improves, namely at the end of the reaction zone. This mode of operation also causes the catalyst, at this important point, decisive for the best possible conversion of the carbon monoxide, to be subject to the least ageing. In the case of adiabatic multi-stage reactors, the end of the pertinent reaction zone is always the most unfavourable point for equilibrium establishment and ageing as the highest temperature always prevails there.

Temperature dependency in regard to byproduct formation can also be quite clearly noted. Effected by this in the first instance are alcohols and, in the case of catalysts with very high percentages aluminium, dimethyl ether also. CO and

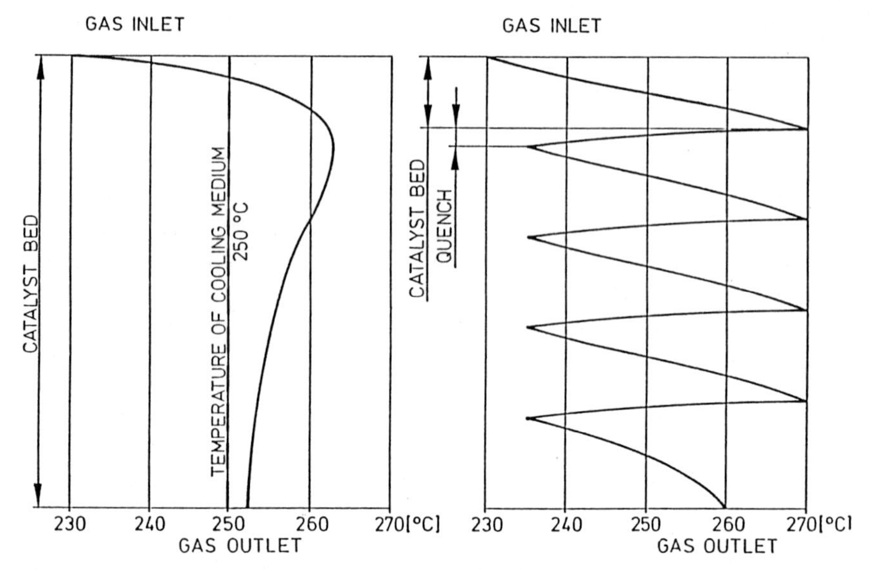

Fig. 3.6. Temperature profiles in methanol reactors. (*Left*) Water cooled tubular reactor; (*right*) quench reactor (saw tooth profile)

CO_2 conversion is also dependent upon the space velocity at which the catalyst is operated. Usually, the reactor feed gas under standard conditions related to the catalyst volume present in the reaction chamber is the measure for space velocity, i.e. m^3 gas/m^3 catalyst. In practice, values between 7 000 and 12 000 are selected. Rising space velocity causes briefer residence time and thus dropping conversion but also permits small catalyst volume, i.e. increasing space time yield, and vice versa. However, there is a maximum in space time yield which cannot be exceeded, even on the space velocity arising further, as the reduction in carbon monoxide conversion is dominant from this point onward.

The ratio recycle gas/synthesis gas, the *recycle ratio*, also influences conversion. At the same time, the size of the equipment in the synthesis loop is also thus determined, as is the power requirement of the recycle compressor. Higher recycle gas ratio causes greater dilution of the reacted gas and higher conversion of the carbon oxides due to the low percentage of methanol in the same. Low recycle ratio reduces the power requirement for the recycle gas compressor and reduces the size of the equipment in the loop. It is the engineer's task to effectively match these sometimes conflicting elements, and their mode of behaviour.

3.3 The Synthesis Loop and Its Various Appearances

3.3.1 The Loop

The previous sections have already shown that the synthesis reactor is indeed the core of methanol synthesis but that in addition to the reactions on the catalyst, a whole series of other procedures take place – heating, cooling, condensing, recompressing – for which suitable equipment must be available. The design of the synthesis plant is almost the same for all processes operating with a gas recycle system. The synthesis gas recycle system is normally termed in brief *methanol loop*. The most important elements of the loop and their arrangement are shown in Fig. 3.7.

The compressed synthesis gas is mixed with the recycle gas and the complete quantity, or some of it, heated in a heat exchanger to the desired reactor inlet temperature. Normally, the temperature of the reactor feed gas is about 220–230°C. In the case of reactors operating with cold quench gas, only some of the reactor feed gas, about 2/3, is preheated. The non preheated gas is fed direct between the individual catalyst layers as cooling gas.

The reaction of H_2 with CO and CO_2 forming methanol (and water) occurs on the catalyst arranged in the synthesis reactor and the reacted gas leaves the reactor at a temperature of about 250–270°C. For the sake of clarity, equipment for removing the reaction heat is not shown in Fig. 3.7. Some of the reacted gas sensitive heat is transferred to the reactor feed gas in the heat exchanger and during this is cooled to the methanol/water dew point, usually even slightly below it. The gas mixture is then cooled to ambient temperature in a cooling stage, often comprising an air cooler and a downstream final cooler operated with cooling water, and during this the methanol and water contained in it largely condensed. Gas and liquid are separated from one another in a separator and the liquid, the

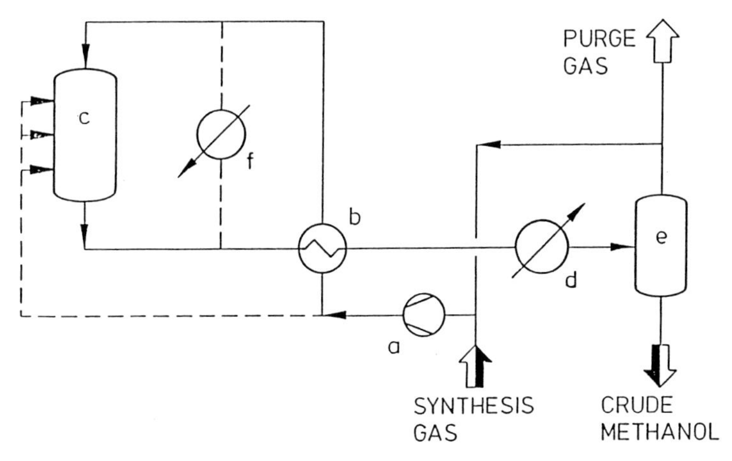

Fig. 3.7. Methanol synthesis loop. (*a*) Recycle compressor; (*b*) heat exchanger; (*c*) reactor; (*d*) cooler/condenser; (*e*) methanol separator; (*f*) start-up heater

crude methanol, is routed for purification or further use. The gas separated from the liquid is returned to the recycle compressor as recycle gas.

Only a small percentage of the inert gases introduced to the synthesis gas, such as methane, nitrogen and argon are dissolved in the crude methanol and leave the loop together with it. To prevent an undesirable concentration of inert gases which would hinder conversion as they reduce the partial pressure of the active gas components H_2, CO and CO_2, a small gas stream must be withdrawn from the loop as purge gas. Synthesis processes where heating the recycle gas in the reactor is not possible, i.e. all processes where the reaction heat is removed outside the reactor or by cold gas quench, require a start heater for starting up, either steam-heated or fired.

In addition to a small quantity of dissolved gas, mainly carbon dioxide, the raw methanol depending upon the CO_2 content of the synthesis gas contains between 3.5 and 12 wt. % water as well as a number of impurities varying with the selectivity of the catalyst used and the CO content of the reactor feed gas. Crude methanol produced from synthesis gas with low CH_4 content recovered from coal generally contains up to 3 500 ppm byproduct in the case of a process operating with a gas loop. Byproduct formation rises to ten times this value in once through processes. Table 3.7 shows a typical crude methanol composition as can be expected on a catalyst with good selectivity and a CO partial pressure of about 6 bar in the reactor feed gas.

Table 3.7. Typical crude methanol composition

Methanol	92.53 % wt.
Water	5.55 % wt.
Dissolved Gases	
CO_2	1.05 % wt.
CO	0.10 % wt.
H_2	0.05 % wt.
CH_4	0.04 % wt.
N_2 + Ar	0.35 % wt.
Light Ends	
Dimethylether + C_5 Hydrocarbons	200 ppm wt.
Acetone	1 ppm wt.
Methylformiate	820 ppm wt.
Others	30 ppm wt.
Heavy Ends	
C_6 - C_{10} Hydrocarbons	30 ppm wt.
Ethanol	1200 ppm wt.
C_3 - C_5 Alcohols	1000 ppm wt.
Others	50 ppm wt.

3.3.2 Methanol Synthesis Using Fixed Bed Catalysts

The two low-pressure methanol processes to which currently more than 90% of the world methanol production is manufactured, the *Imperial Chemical Industries*, Ltd. (ICI) one and the LURGI one, differ considerably in the reactor section.

In the *ICI Process* of the first generation, a quench reactor as shown in Fig. 3.8 is used [3.20]. Figure 3.9 shows a typical ICI loop. Some of the reactor feed gas is preheated to almost reaction temperature in a heat exchanger and enters the first catalyst layer in the reactor from above whereas a smaller portion of the feed gas is applied as quench gas without preheating. *Quench gas* is fed into the reactor downstream of each catalyst layer, with the exception of the bottommost, through lozenge distributors. As is to be seen in Fig. 3.8, these cage-like structures are annular causing largely uniform gas distribution and mixing across the reactor cross section, allowing simple charging of the reactor with catalyst (from above) and emptying it (downward). The quench gas is dosed by temperature controllers provided with pulses in each case from the bottom end of the catalyst layers.

The temperature gradient in the reactor, shown across the height of the overall catalyst chamber, gives the *saw tooth diagram* typical for quench reactors, as already shown in Fig. 3.6. Some of the reacted gas flows through the heat exchanger after it leaves the reactor where it heats the reactor feed. The sensible heat of the other portion is used for preheating feed water, as shown in Fig. 3.9, or to produce low-pressure steam. The gas cooled to about the methanol/water dew

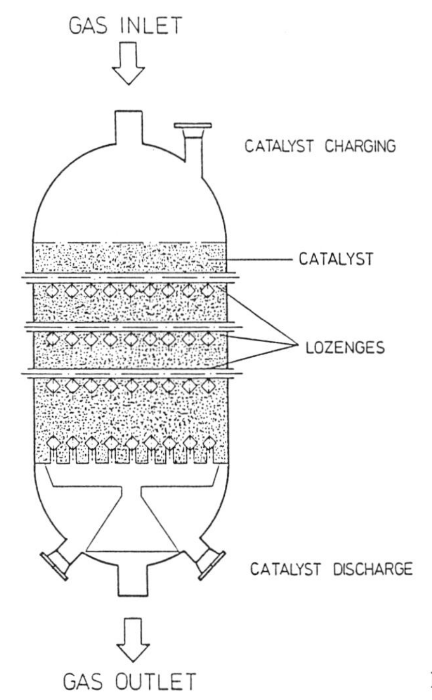

Fig. 3.8. ICI quench reactor

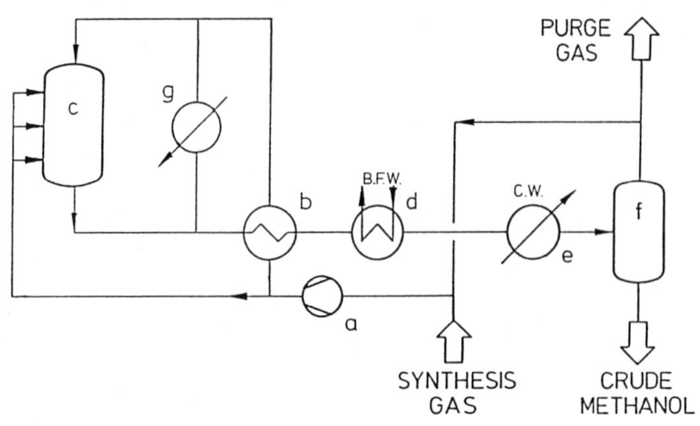

Fig. 3.9. ICI methanol synthesis loop

point is then further cooled by an air cooler, the majority of the methanol and water condensing. The gas is usually finally cooled in a water cooler downstream of the air cooler. Thereafter, the gas/liquid mixture enters the separator where the crude methanol is separated from the recycle gas and routed through a flash tank to the methanol distillation or to a crude methanol tank.

The recycle gas is mixed with the makeup gas and again compressed to the loop input pressure with the aid of the recycle gas compressor. The recycle gas compressor is designed as a single-stage turbo-compressor, usually driven by a steam turbine. The purge gas is withdrawn from the loop pressure-controlled maintaining the suction pressure at the recycle compressor constant downstream of the methanol separator and upstream of the point at which the makeup gas is added. A start heater serves for starting up the plant which in general is suitable for steam heating and is arranged in the reactor gas line between heat exchanger and reactor. The inlet temperature into the reactor can be maintained high enough, even in the case of failures, by this start heater to avoid the reactor *extinguishing* which is one of the possible evils of a quench reactor.

Plants of this type normally operate in the pressure range between 50 and 100 bar and with recycle/makeup gas ratios of 5 to 7. In the case of very high capacities, this ratio is occasionally reduced to 3.5, a somewhat poorer carbon oxide yield being accepted.

In such cases, the synthesis gas reactor is then equipped with an increased number of quench gas controllers in order to maintain the higher temperature rise in the individual catalyst layers to be expected due to the weaker dilution in limits. The pressure drop across the overall loop is 6–10 bar. The largest plants in operation have methanol reactors with an individual capacity up to 2 500 tpd. Since 1976, a plant with a capacity of 50 tpd has been in operation in South Africa which is fed with synthesis gas from a coal gasification unit operating to the *Koppers* process.

A certain limitation in regard to its maximum size is dictated for the multi-stage quench reactor in view of the pressure drop at the catalyst [3.21]. For that

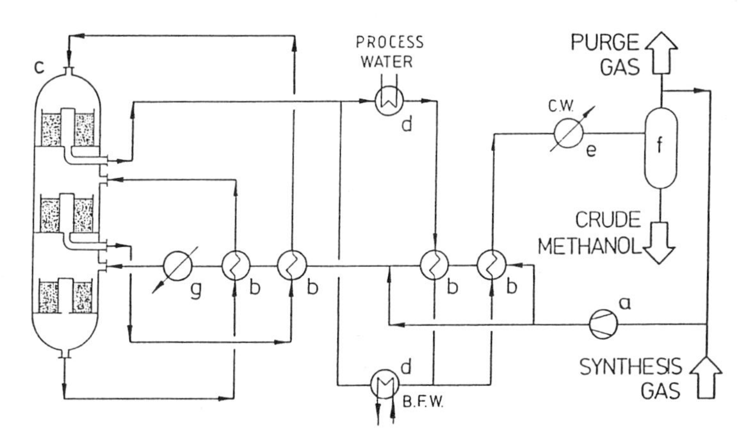

Fig. 3.10. Methanol synthesis with axial/radial flow reactor. (*a*) Recycle compressor; (*b*) heat exchanger; (*c*) reactor; (*d*) BFW preheater; (*e*) cooler/ condenser; (*f*) separator; (*g*) start-up heater

reason, another type of reactor was proposed for the ICI process to cope with capacities up to 5 000 tpd which were considered for gigantic plants to produce fuel grade methanol.

The ammonia technology of *Ammonia Casale*, in which an axial/radial flow converter with intercooling is applied, was taken for this purpose. Figure 3.10 shows this reactor in a methanol loop, only little changed otherwise from the original ICI design. Due to the intercooling effected by the reactor feed gas, a number of additional heat exchangers are required. The reactor feed gas is brought up to reaction temperature in these heat exchangers and then entered the first catalyst chamber in the bottom section of the reactor. The gas flows through the annular arranged catalyst, largely from the outside to the inside, and leaves the catalyst chamber again through a central tube. After appropriate cooling in an intermediate cooler, the gas, already reacted in part, then enters the next catalyst chamber. This procedure can be repeated two or three times and finally, the reacted gas leaves the reactor from the top catalyst chamber. The surplus heat remaining from the reaction is utilized to preheat feed water or generate low-pressure steam, as was already the case in the ICI loop of the first generation.

In comparison with the reactor with axial flow, the axial/radial flow converter offers a means of operating with low pressure drop across the catalyst despite its slender design. Although this advantage is partially eliminated by the intermediate heat exchanger, the pressure drop could be reduced using the arrangement shown in Fig. 3.10 to about 5.5 bar – against about 7 bar in a loop with quench reactor – for the concept of a plant to produce 2 500 tpd methanol. Due to poorer space utilization, the weight of an axial/radial flow converter is considerably higher than that of a quench reactor of the same capacity despite a possible reduction of the catalyst volume of about 20 % due to the differing reaction control and despite the slender design. Temperature control of the axial/radial flow converter can be maintained more stable by a simpler means than that of a

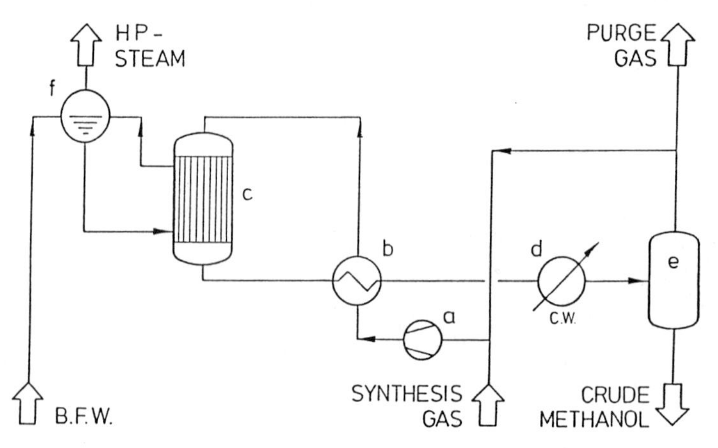

Fig. 3.11. *LURGI* Methanol synthesis loop. (*a*) Recycle compressor; (*b*) heat exchanger; (*c*) reactor; (*d*) coller/condenser; (*e*) separator

quench reactor in which a temperature change can only be effected by modifiying the quench gas streams. A catalyst change in an axial/radial flow converter takes considerably more effort than in a quench reactor equipped with lozenge distributors as each reactor chamber must be individually filled or emptied.

Latest proposals from ICI include a water cooled reactor type as well as a tubular gas cooled reactor similar to the type that has been used since some decades for high temperature CO shift conversion (Sect. 2.4, Fig. 2.21)

The *LURGI Low-pressure Process*, which was developed to suitability for operation, as was the ICI process, in the second half of the sixties, used a *tubular reactor* cooled with boiling water from the beginning of the development work and still does this today. Figure 3.11 shows the LURGI loop with the integrated reactor. The reactor feed gas is preheated to approximately reaction temperature in the heat exchanger and enters the reactor from the top, flows through its tubes filled with catalyst from top to bottom, the reaction of the carbon oxides with hydrogen, formerly described, occurring during this. The reacted gas leaves the reactor at the bottom, flows through the heat exchanger, where it preheats the reactor feed gas, and then passes into the condensation/cooling section usually comprising an air cooler and a cooling water-cooled final cooler. Here, it is cooled to ambient temperature, methanol water and other condensable gas components being condensed. The crude methanol is separated from the recycle gas in the downstream separator and passes through a flash tank to the distillation unit (Chap. 4.).

The recycled gas is routed to the recycle compressor after the purge gas is withdrawn through a pressure controller. Here it is brought up to the makeup gas pressure and mixed with this gas. The makeup gas can also be mixed with the recycle gas on the suction side of the recycle gas compressor. The feed point is stipulated appropriate to the pressure increments of the synthesis gas compressor and possible heat recovery in the condensation area of the loop. Generally, common compression of makeup and recycle gas in the recycle compressor effects a

Methanol reactor with steam drum and heat exchanger

slight saving in the sum of the energy requirement for makeup and recycle gas compressors. As opposed to this, mixing the makeup gas leaving the makeup gas compressor at elevated temperature offers the means in the methanol condensation area, which now and then starts at temperatures up to 130°C, of still using heat from the reacted gas, e.g. for preheating feed water.

The reaction heat is removed by a boiling water system with natural circulation. Water at almost boiling temperature flows to the bottom section of the reactor from the steam drum arranged above the reactor. The heat is transferred from the tubes where the reaction takes place, to the water and during this, some of it evaporates. This causes the natural circulation. The water/steam mixture leaves the water section of the reactor again at the top and rises back to the steam drum where the generated steam is separated from the non-evaporated water. The control of the boiling water temperature and thus of the temperature gradient in the reaction zone also is effected by the steam pressure control in the steam line leaving the steam drum. This method permits simple and very constant control of the temperature prevailing at the catalyst as the temperature rise of saturated steam in the range between 240 and 260°C progresses very flatly above the pressure. A pressure change from 40 to 42 bar, for example, would only cause a change in temperature of about 2.5°C. Due to a combination between the supply of feed water and boiling water, the feed water being supplied at low temperature, steam can be generated with pressures up to 60 bar and despite this a temperature in the bottom section of the reactor sets, which is only about 245°C, so that at the end of the catalyst bed, a favourable equilibrium temperature is obtained [3.22].

Steam is fed through an injector into the boiling water circulation system for starting of the plant or in the case of lengthy interruptions to operation when a quick synthesis startup is required. This method causes uniform heating of the catalyst and also allows easy maintenance of catalyst heat during short shutdown periods. If the synthesis is supplied with synthesis gas with low inert content and a pressure of at least 25 bar which is the case in most coal gasification systems, the quantity of steam generated in the synthesis reactor with appropriate superheating is sufficient to cover the power requirement for the synthesis gas compressor, the recycle gas compressor and as exhaust steam the heat requirement for distillation.

Plants built to the LURGI process operate in the pressure range of between 50 and 100 bar and with recycle gas/makeup gas ratios of 3.0 to 4.0. Even in the case of makeup gases with high CO content, operating at low recycle gas/makeup gas ratios is possible as the high speed of the CO conversion does not lead to local overheating in the catalyst due to the rapid transfer of heat from the catalyst to the boiling water. The pressure drop across the overall synthesis loop is only 3.5 to 4 bar largely attributable to the fact that there are no control or throttle elements in the whole gas route.

The largest reactor in operation has a potential capacity of max. 1 800 tpd and until recently was operated with makeup gas from the partial oxidation of vacuum residue which is almost identical with the synthesis gas from coal.The plant is now supplied with synthesis gas from a coal gasification unit. The first synthesis for processing a gas with a high CO content at a capacity of 600 tpd has been operating in Germany since 1973 and a plant with the same type of gas at a capacity of 2000 tpd in the USA since 1980. Since 1983, a 450 tpd plant has been in operation in the USA which is supplied with makeup gas from a TEXACO coal gasification.

Mitsubishi Gas Chemicals put their *MGC Methanol Synthesis* unit into operation in Japan during the first half of the seventies. This plant was originally designed for a pressure of 150 bar but was then operated with satisfactory results at pressures less than 100 bar. The process is usually not termed a low-presure process but a medium-pressure one. The concept of the synthesis loop scarcely differs from that for other methanol processes operating with quench reactors; in fact, a quench reactor was also used by MGC at the beginning. Later the design was modified to intercooling between each reaction stage, low-pressure steam being generated at the same time as shown in Fig. 3.3 [3.23].

Today this process is offered in the pressure range between 50 and 200 bar and operates at temperatures between 235 and 270°C. Three plants are in operation, two indeed at MGC. These two plants use makeup gas produced from natural gas. A third plant is operating in the Arab world.

Messrs. *Haldor Topsoe* A/S, Denmark, uses a radial flow reactor of its own development [3.24] and similar to the MGC process, this is more a *medium-pressure* than a low-pressure process. Several of these reactors are arranged in series in the synthesis loop, each reactor representing one reaction stage. The necessary gas cooling at the outlet of the reactor before entering the next is effected by heat transfer to feed water or generating low-pressure steam. For plants of this type, the design provides for pressures up to 150 bar and a temperature range of 200 to max. 310°C is given. It is not known if the maximum temperature can also be realized in continuous operation without catalyst damage. To date nothing is known on the construction of plants to this process.

Messrs. *Linde AG*, Germany, has being propagating its *Variobar*$^{(R)}$ process, a boiling water reactor highly similar to the methanol reactor type developed by LURGI, since 1982. The catalyst is arranged outside the spiral-shaped boiling water tubes. In a similar manner to the LURGI reactor, the reactor introduced by *Linde* also permits 40 bar steam to be produced. The synthesis loop is almost identical with that for the LURGI process. Process pressures of 50–150 bar are stated and it is to be supposed that the preferred temperature range is between 240 and 270°C. Dissimilarly to the four process owners mentioned previously, Messrs. *Linde* does not offer proprietary catalyst. To date, one commercial reactor is in operation.

An interesting route for producing methanol is described by *Chem Systems Inc.*, USA, under the sponsorship of the *Electric Power Research Institute (EPRI)*. It has developed a process in which the synthesis catalysts is suspended as slurry in a liquid hydrocarbon. An initial design introduced in about 1976 operates with fine-grained catalyst in a fluidizing bed, the liquid hydrocarbon being used to agitate the catalyst as well as remove the reaction heat. Figure 3.12 shows a slightly simplified loop as proposed by *Chem Systems*. Synthesis gas is preheated,

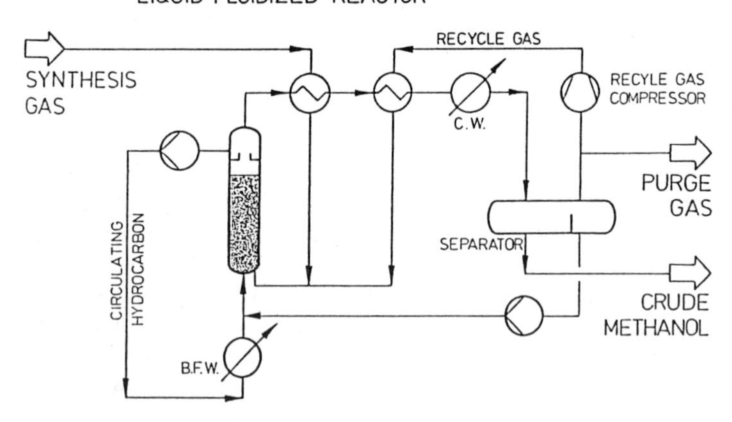

Fig. 3.12. *Chem Systems* three phase methanol process

mixed with a stream of recycle gas, also preheated, and introduced together with the liquid hydrocarbon into the fluidized bed reactor from below. Here, reaction takes place on the agitated catalyst.

Gas, liquid and catalyst are separated in the top section of the reactor. The gas also containing the methanol formed as vapour is routed to two heat exchangers where it is used to preheat the gases flowing to the reactor and is cooled in a cooler, the methanol (and water) condensing. After the methanol/water mixture is separated, the remaining gas is recycled to the reaction after a certain quantity of purge gas is discharged. The liquid, the hydrocarbon, separated in the top section of the reactor, is pumped through a waste heat boiler, cooled there to reactor inlet temperature and during this medium-pressure steam is generated. The percentage of hydrocarbon entrained in the gas leaving the reactor condensed together with the methanol is separated from the methanol by phase separation in the methanol separator as the hydrocarbon does not mix with it and returned to the hydrocarbon recycle system.

It is stated [3.25] that the process can operate between 35 and 100 bar and at temperatures of 200–280°C. In view of the low recycle/makeup gas ratio, methanol contents in the reacted gas up to 14.5% ought to be reached.

3.3.3 Other Methanol Syntheses

Newer developments by *Chem Systems* are aimed at using a *liquid-entrained* catalyst reactor instead of a fluidized bed. The necessary catalyst is supplied in pulverized form and also agitated in a hydrocarbon in this variant. Aliphatic mineral oils in the range C_{14} to C_{21} appear to be particularly suitable as carrier liquid [3.26]. This variant ought also to attain high conversions applying a single-pass mode of operation which however, of course, is limited by the natural equilibrium. The desulfurized synthesis gas is preheated in a heat exchanger to reaction temperature and passes through the reactor through which the catalyst/mineral oil mixture flows either in cocurrent or in countercurrent. Gas and catalyst/mineral oil mixture are discharged separately from the reactor, condensation and separation of the methanol from the offgas as well as recycling the catalyst/mineral oil mixture to the reactor are effected in the known way. Here, the expression *offgas* instead of the word purge gas, otherwise used, is intentionally applied as due to the single-pass operation, the active non-converted gas components are many times higher than they are in the case of processes operating with gas recycling.

Here, one can speak rather more of coproduction of methanol and fuel gas. Equilibrium adjustment assumed, only 26.7% of the CO from a synthesis gas containing 50% CO and 50% H_2 is converted at 100 bar and 250°C. Thus 5 240 m^3 synthesis gas would have to be made available to produce 1 ton of methanol. Of this quantity, about 2 100 m^3 would be converted to methanol and the remaining 3 140 m^3 would be obtained as fuel gas with about 60% CO and 40% H_2. In other words, only about 32.5% of the heat carried in the reactor in the synthesis gas is contained in the methanol.

Chem Systems operate with methanol catalyst attainable on the market but do, however, carry out their own development work also. This is mainly directed at the application of the methanol produced as combustion fuel (not as motor fuel) as this application is definitely in the foreground of the *Chem Systems* endeavors regarding process development. Crude methanol produced in the reactor with the liquid-entrained catalyst is particularly little suited to manufacturing pure methanol due to its high percentage of byproducts as treatment of this methanol is technically difficult and costly.

Recent other developments range from use of homogeneous, liquid catalysts to direct oxidation of methane.

Brookhaven National Laboratory, USA claims a process in which methanol is produced from carbon monoxide and hydrogen over a liquid phase homogeneous catalyst at some 100°C and pressures between 10 and 15 bar. Carbon dioxide – even in small amounts – is detrimental to catalyst lifetime, CO conversion of 94% is expected. Selectivity of the catalyst seems to be extremely high.

University of Manitoba, Canada, has published [3.27] results of laboratory tests on direct conversion of methane to methanol. Pressures applied range from 30 to 70 bar, temperatures from 300 to 430°C. At space velocities comparable to those of commercialized processes, the conversion rate is between 4 and 10% of the natural gas feed only, and methanol selectivity is between 70 and 90%, the balance being mainly CO_2. As oxidant, pure oxigen has been proposed.

Twente University of Technology, Netherlands, proposes two novel systems, one being a *Solid Trickle Flow Reactor*, the other one a *Reactor System with Interstage Product Removal* [3.28].

Although these proposals exhibit interesting new ideas, it still has to be demonstrated whether they can play a competitive role against established technologies considering economical and safety aspects.

3.4 Production of Mixed Alcohols

3.4.1 Why Mixed Alcohols and Which Ones

The most important feature of the processes described to date for manufacturing chemical methanol is a reactor inlet gas exhibiting an H_2/CO ratio of more than 5. The endeavor in all these processes is to produce as few impurities in the crude methanol as possible. In the case of the processes aimed at producing methanol for combustion fuel purposes, one operates at considerable lower H_2/CO ratios and higher percentages of higher alcohols or hydrocarbons are not viewed unfavourably. They do, indeed, slightly reduce the efficiency of the CO conversion but, on the other hand also cause an increase in the crude methanol calorific value.

The most recent past has just indicated that the known difficulties occurring on mixing methanol with gasoline can be eliminated in that a mixture of methanol and higher alcohols is used. This increases the octane number, prevents water demixing and effects a vapour pressure drop with the result that less C_4 hydrocarbons are forced out of the natural motor fuel. The following requirements are placed upon these mixed alcohols:

- A content of higher alcohols in the C_2 to C_6 range of 30 to 50 wt. %
- Highest possible wt. % of C_3 to C_5 alcohols as large quantities of C_2 can lead to hot engine problems and the alcohols with 6 C-atoms and more, no longer contribute to the same degree as the lower alcohols to increasing the octane number.
- The water content shall be below 0.5 wt. %.
- The mixture shall not contain any corrosive components or such leading to corrosive combustion products.

In addition, it is naturally demanded of a process to produce mixed alcohols that it operates at the lowest possible pressures and temperatures – similar to the modern methanol processes – and with economic CO conversions, in one word economically.

3.4.2 Various Types of Mixed Alcohol Processes

Since 1980, a number of patents have been applied for in this field [3.29–31] and a number of publications have appeared [3.32,33]. The processes belonging to *Snam Progetti* S.p.A. and LURGI have progressed to the technically realizable stage. Both processes operate with H_2/CO ratios of 0.5 to 2.5 at the reactor inlet and the pressure range is between 50 and 150 bar. The temperature range and the catalysts applied clearly differ just as the water content in the raw product also does. This results in considerable differences in distillation (Sect. 4.3).

The *Institute Français du Petrol* also reports on its own developments [3.30] which have apparently not yet progressed far enough to permit an optimum catalyst to be finally specified.

The development of the *MAS process* was initially worked on by *Snam* alone but later the Italian ANIC and the Danish *Haldor Topsoe* A/S also participated. The synthesis loop is almost identical to that of a normal methanol synthesis and a reactor with direct or indirect cooling can be applied. The newly developed catalyst contains chrome and zinc in a ratio of about 3 : 1 as well as low quantities of about 2.5 wt. % alkali, preferably calcium or potassium. 50–100 bar bar are given as a possible pressure range but here one appears to prefer the high pressures in view of the better CO conversion. The reaction proceeds in a temperature range between 350 and 450°C. A synthesis gas is used as makeup gas with an H_2/CO ratio of less than 2.0 and exhibiting a very low CO_2 content. In the synthesis loop in which a recycle gas/makeup gas ratio in the order of 10 is operated, the H_2/CO ratio rises and is somewhat more than 2 in the reactor inlet gas. A reactor outlet gas is obtained containing in addition to 57 wt. % methanol about 25 wt. % C_2–C_8 alcohols and 17 wt. % water as condensable substances. The major percentage of the higher alcohols is butanol at about 45 wt. %. The alcohol mixtures must be distilled in several steps for use as octane booster in natural gasoline (see Chap. 4).

The LURGI *Octamix Process* developed in the years 1978 to 1983 also applies a synthesis loop highly similar to that for the LURGI low-pressure methanol process. As in the case of this process, the tubular reactor is also used to produce fuel methanol. A modified variant of the LURGI methanol catalyst is used operating optimally at a temperature range between 260 and 290°C and at pressures below 100 bar.

Differing from other processes to manufacture alcohol mixtures, some reducing the CO_2 content in the makeup gas and some operating without any CO_2 being removed, the CO_2 content in the reactor gas, i.e. inside the synthesis loop, is reduced to values below 1 vol. % in the Octamix process. A H_2/CO ratio in the reactor inlet gas of 0.5 to 1.0 prevails at a recycle gas/makeup gas ratio of about 5 with a makeup gas having a H_2/CO ratio below 2.0. This operation together with the low operating temperature leads to the water occurring during the formation of alcohols higher than methanol to the equation

$$n\mathrm{CO} + 2n\mathrm{H_2} = \mathrm{C}_n\mathrm{H}_{(2n+1)}\mathrm{OH} + (n-1)\mathrm{H_2O}$$

being largely consumed in the reaction zone to convert CO to CO_2. A raw product is obtained with such a low water content that it does not disturb in mixture with natural gasoline and which after the dissolved gases and low boilers are separated has approximately the following composition:

Methanol	60.5 wt.%
Ethanol	6.0 wt.%
Propanol	4.0 wt.%
Butanol	7.0 wt.%
Pentanol	7.0 wt.%
Hexanol$^+$	6.0 wt.%
Water	0.5 wt.%

The propanol is available largely as $n - C_3$, the butanol mainly as $i - C_4$. The content of higher alcohols can be simply increased by distillation due to the low water content and here top methanol of grade AA can be produced .

A methanol process intended exclusively to produce fuel methanol was also launched by Messrs. *Wentworth Brothers*, Inc., Cincinatti, in 1982 [3.34]; it is based on experience gained with the high-pressure process which has become classical under the name *Vulcan Cincinnati*. Contrary to the rules of thermodynamics, the attempt to operate with low temperatures at the end of the reaction and thus under favourable equilibrium conditions was not always consequently made. As opposed to this one appears, at least in intermediate sections of reaction control, to aim at as high an end temperature as possible in order to be able to utilize the effective part of the reaction heat at the highest possible temperature level to generate high-pressure steam and probably also to improve the conversion of CO_2.

Wentworth Brothers Inc., do not, as do other modern processes operating with intermediate cooling, remove heat from between all reaction zones (catalyst layers) arranged one after the other to maintain as low an operating temperature as possible in the following zone. It is much more the case of using many reactors, at least 3, each of which is filled with a different catalyst.The reacted gas from the first reactor enters the next reactor without intermediate cooling, which is filled with a catalyst matched to this higher temperature and so on. At the outlet of the third reactor, the partially reacted gas then flows through a waste heat boiler generating high-pressure steam. Finally, this gas then flows through two or three further reactors, intermediate cooling again being practiced upstream of the last reactor.

In the lower temperature range – from approximately 220 to 260°C – copper-rich catalysts are used whereas in the temperature range up to about 360°C, low-copper, high-zinc catalysts are used. As a temperature higher than 380°C could not be permitted even on applying the old zinc/chrome catalysts, one can assume that operation is carried out with a recycle gas/feed gas ratio of about 4.5 to limit a temperature rise from 220 to about 380°C. If one continues such deliberations under this prerequisite, one notes that about 70 % of the reaction heat can be recovered.

In the case of a gas containing no, or only very little, CO_2 that would mean that about 900 kg high-pressure steam per ton of methanol with a pressure up to max. 40 bar can be generated as against about 1 100 kg with tubular reactors under analogous conditions.

To date, there are no plants to this modified process existent and it is not known whether the further development of the copper and zinc catalysts, in fact known, has led to the required catalyst assortment applying which there can be optimum operation over a wide temperature range.

4. How to Obtain Pure Methanol

4.1 Why Refining Methanol

As described already in Sect. 3.3.1 above, raw methanol contains not only water stemming essentially from the reaction of CO_2 with H_2 and, to a lesser extent, from the formation of dimethyl ether and higher alcohols according to the reactions

$$2CH_3OH \rightleftarrows CH_3OCH_3 + H_2O$$

$$CH_3OH + nCO + 2nH_2 \rightleftarrows CH_3(CH_2)_nOH + H_2O$$

but also a number of substances whose boiling points are in some instances lower and in some higher than that of methanol. They are usually termed low boilers and high boilers or light ends and tails and include mainly

- dissolved gases from the synthesis loop
- dimethyl ether
- methyl formiate
- acetone (in very small quantities)
- C_5 to C_{10} hydrocarbons (traces also up to C_{30})
- ethanol
- higher alcohols up to C_5 (and traces of C_5^+).

The raw methanol may also contain substances which were not formed in the synthesis reactions but come from the synthesis gas or from the equipment (such as iron).

Unlike the raw methanol produced at high pressures over a zinc-oxide/chromium-oxide catalyst in the past, the reaction of copper catalysts produces hardly any aldehydes or ketones. Although the fact that methyl formiate or methyl acetate can be detected, suggests that minor quantities of formic acid and acetic acid are formed as well, corrosion by formic acid in the raw methanol condensation section, which had been so much dreaded in the high-pressure synthesis processes, has not been described for low-pressure syntheses.

The objective of methanol purification is to remove these impurities until a marketable product is obtained. Raw methanol produced over copper catalysts is today purified exclusively by way of distillation, except for a few cases in which traces of impurities (such as amines) have to be eliminated by ion exchange.

Whenever 20 years ago the discussion turned to the subject of synthetic methanol, the reference was normally to pure methanol of a quality as defined by US Federal Grade Specification O-M 232 which in its revision of June 5, 1975 still defines even today which impurities pure methanol may contain. Two different qualities – Grade *A* and Grade *AA* – are listed in Table 4.1. However, some methanol processing companies go even beyond the Grade AA require-

Table 4.1. Requirements for grade A and grade AA methanol

Characteristics	Grade A	Grade AA
Acetone & Aldehydes % by wt. max.	0.003	0.003
Acetone % by wt. max.	n.a.	0.002
Ethanol % by wt. max.	n.a.	0.001
Acidity as Acetic Acid % by wt. max.	0.003	0.003
Alkalinity as NH_3 % by wt. max.	0.003	0.003
Appearance	Clear & colorless	same
Carbonizable Substances	No discoloration	same
Color	Not darker than color standard No. 5 of ASTM Platinum Cobalt Scale	same
Distillation Range	Not more than 1 °C and shall include 64.6 °C +/- 0.1 °C at 760 mm	same
Hydrocarbons	No cloudiness or opalenscence	same
Specific Gravity	0.7928 at 20 °/20 °C	same
Percent Methanol by wt. min.	99.85	99.85
Non Volatile Content % by wt. max.	0.001	0.001
Odor	Characteristic, non-residual	same
Water % by wt.	0.15	0.10

ments today, for instance by calling for water contents of less than 50 ppm, while others such as the manufacturers of acrylates are satisfied with ethanol contents of less than 100 ppm since less than 10 ppm of ethanol are required only for dimethyl phthalate production. The lastest ASTM Specification D 1152-77 for methyl alcohol makes no reference to ethanol and thus obviously leaves the specification of ethanol contents to an accord between producers and consumers. Some formaldehyde processes can do with methanol qualities which do not even meet Grade A requirements; especially the water content may often be considerably higher.

New methanol applications developed since the end of the seventies tolerate even a number of intermediate qualities between Grade AA and raw methanol. The MTG (Methanol-to-Gasoline) process developed by *Mobil Research and Development* for the production of gasoline, for instance, uses untreated raw methanol from which only permanent gases have been removed by letting it down to atmospheric pressure. It must only be ensured that it does not contain any substances – essentially such basic components as amines – which may damage the zeolite. Nor are any quality requirements imposed on methanol which is used as a burner fuel (not as a motor fuel); it must only be storable, i.e. both permanent gases and components with very high vapour pressures – normally only dimethyl ether – have to be expelled to a residual content ensuring that the methanol does not release significant quantities of gases upon storage. If the water content of raw methanol and the transportation costs are high and waste heat of more than 115 °C is available at low cost, it may be economically rewarding to remove the water to residual contents of 0.5 to 1 wt. % in the fuel methanol.

However, if the effluent quality is to be such that no further treatment is required, this operation consumes some 1.3 GJ per ton of methanol or 600 kg of steam at a minimum pressure of 1.5 bar. Methanol qualities which are to be used as blending components for motor fuels, on the other hand, have to conform to more exacting demands, and their water contents have to be reduced to some 0.1 to 0.3 wt. %.

4.2 Basic Methanol Distillation Systems

The requirements described in the previous chapter have led to the development of two different basic methanol distillation systems which are shown in Fig. 4.1. In the *single column system*, the low boilers are expelled overhead, while the product methanol from the column top section is withdrawn, depending on the desired quality from the column top above the reflux inlet and the process water is discharged from the column bottom. The high boilers may be withdrawn either together with the process water or through a side outlet below the crude methanol inlet. If a side outlet for high boilers is installed, the methanol losses through the process water can be reduced and – if the column is appropriately designed – the process water will be sufficiently pure to meet authority requirements in most industrialized countries.

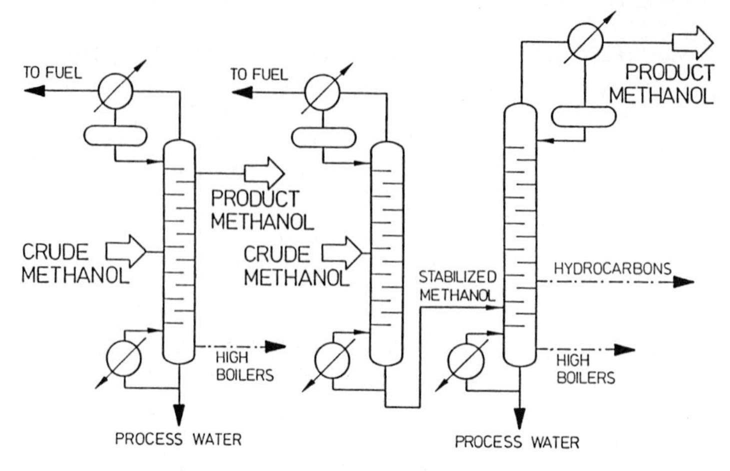

Fig. 4.1. Single column (*left*) and two-column methanol distillation system

With the single column system it is not possible to produce pure methanol of Grade A or even AA under economically justifiable conditions, and it can therefore be used only in plants which are scheduled to turn out methanol for burner fuel or motor fuel applications.

A *two-column system* always consists of a light ends column, which is sometimes also termed an extraction column, and a refining column. The former is used to expel the light ends overhead, or sometimes only the dimethyl ether, which can be conveniently marketed. If this is the case, the other low boilers are withdrawn from a side outlet above the inlet. Methanol, water and high boilers are withdrawn from the bottoms and fed to the refining column in which as a rule the pure methanol is obtained at the column top, the high boilers below the feed nozzle, and the water in the column bottoms.

In spite of its name, the two column system frequently consists of more than two columns. In order to save energy, the refining column is often split up into a first stage operating at elevated pressure and a second atmospheric stage. The methanol vapours from the top of the pressurized stage are condensed in the reboiler of the atmospheric stage and used to heat it. The number of columns has thus increased already to three, and a fourth, normally small, column may even be required to expel methanol and high boilers – often separately – from the process water in the bottoms of the refining column. This is done to keep methanol losses to a minimum and to obtain a waste water which may be discharged to the sewer without further treatment. However, since syngas produced from coal normally contains only little CO_2, 4–6 wt. % as a rule, such plants are particularly suitable for using process water from the bottoms of the refining column without further treatment to generate process steam for coal gasification. Although in this way the raw methanol yield is slightly reduced as a consequence of apparent methanol losses, the quality of pure methanol produced per unit of coal remains practically unchanged as the alcohols recycled to the gasification section together with the process water are broken down there into H_2, CO and CO_2.

Methanol distillation – two columns system

4.3 Production of Grade AA Methanol

The flowsheet in Fig. 4.2 shows a typical distillation system which may be used to describe the process reactions and main characteristics of the equipment used. In addition to dimethyl ether, methyl formiate, acetone and hydrocarbons up to about C_8, the low boilers in a broader sense include also that part of the gases dissolved in the raw methanol which is not expelled when this is flashed into an interim storage tank. This flashing operation is normally performed at pressures between 5–10 bar in order to keep methanol losses down, but frequently also in order to ensure that the raw methanol can be transferred to the light ends column without using a pump.

Whenever the occurence of amines has to be expected – for instance due to the presence of nitrous oxides or ammonia in the synthesis gas – a cation exchanger is normally provided in the predegassed raw methanol. Raw methanol is fed to the bottom half – usually between trays 8 and 10 counting from the top – of the prerun column which is equipped with a total of some 40 trays. Valve trays are normally used as they are characterized by a much better partial load behaviour than the cheaper sieve trays. The column is heated by a reboiler

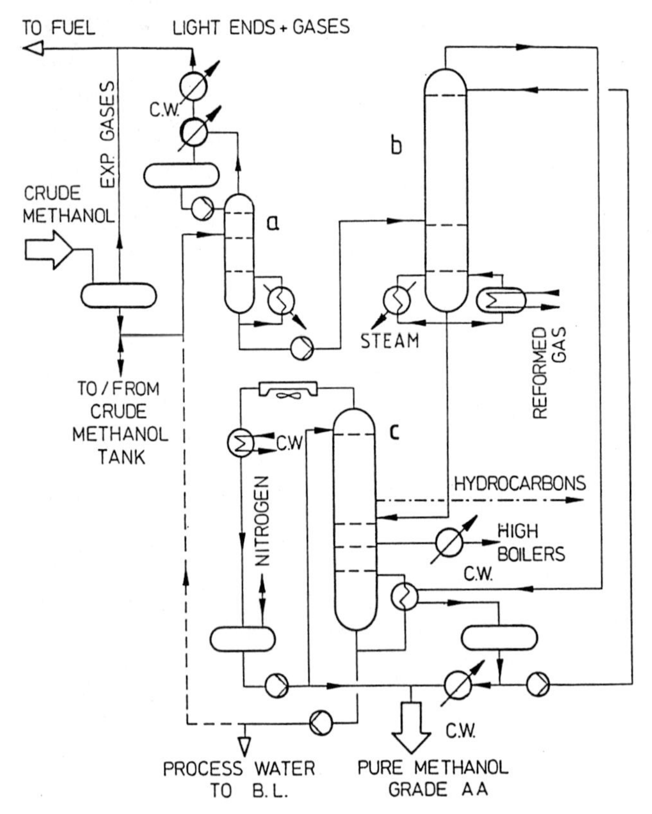

Fig. 4.2. Grade AA methanol distillation process. (*a*) Light ends column; (*b*) elevated pressure column; (*c*) atmospheric column

installed at the bottom which in most cases feeds on low-pressure steam although sometimes waste heat from the raw gas may be used as well. The latter may have its merits above all if the steam dew point of the raw gas lies in a temperature range in which the condensation heat can no longer be used for other purposes such as preheating boiler feedwater. The vapours rising to the column top as the liquid in the bottom is boiled contain permanent gases, low boilers, methanol and steam. Water, methanol, and low boilers in the upper boiling range are condensed at the column top and refluxed while the dissolved gases and the uncondensed low boilers are withdrawn from the system. Condensation is normally performed in two stages. This ensures that a lower temperature differential between the condensate and the cooling water can be reached without unjustifiable expenditure, and methanol losses from the low boilers column can be kept down.

Those low boilers which are not liquefied in the overhead condenser include mainly dimethyl ether, methyl formiate and low boiling hydrocarbons. The water content of the raw methanol influences the other escort substances of the raw methanol and changes their activity coefficients. Column operation is determined by the so-called key components. The aim is to keep those hydrocarbons out of the refining column, whose volatility is near that of water-free methanol or higher as they would either get into the pure methanol, at least partly, or – if withdrawn through side outlets – would lead to methanol losses. Some of the process water from the refining column is therefore added to the raw methanol whenever its water content is low. N-nonane or n-decane, depending on their percentages in the raw methanol and on its water content, are selected as key components for hydrocarbon removal.

Figure 4.3 shows vapour/liquid equilibria (y versus x) in mol fractions for mixtures of methanol and C_6 to C_{10} hydrocarbons with the points of inflection

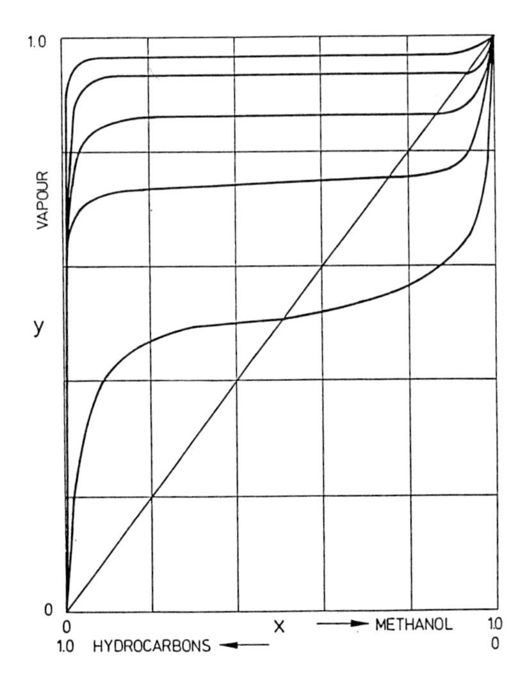

Fig. 4.3. Vapour/liquid equilibrium of a methanol/hydrocarbons mixture

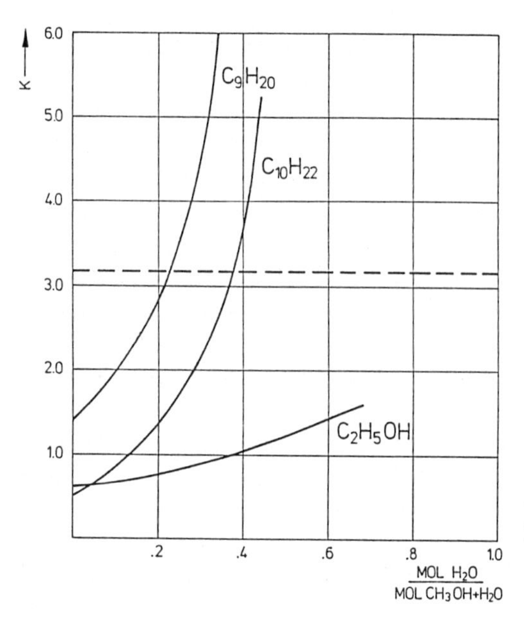

Fig. 4.4. Influence of water content in methanol upon activity coefficients of C_9- and C_{10}-hydrocarbons and of ethanol

lying on the diagonal. The curves indicate that, for a mixture consisting of almost 100 % methanol with only a few ppm of C_9 or C_{10} hydrocarbons, if the vapour/liquid ratio is sufficient, the C_9 would tend to rise from the raw methanol inlet towards the column top, whereas the C_{10}, provided that the vapour/liquid ratio is not too high, would get into the bottoms of the prerun column and would then have to be eliminated in the pure methanol column. As mentioned above, the addition of water to such mixes changes the equilibrium constants ($K = y/x$) of the hydrocarbons. What effects are produced by the presence of water is shown in Fig. 4.4 in which the K-values for C_9 and C_{10} hydrocarbons are plotted against the water content of the methanol. The data apply to the normal operating range of a prerun column, i.e. for approximately 2 bar and 85°C. An accurate calculation of these values has become possible only in recent years with the aid of modern computer programs by which the equilibria for binary mixtures, which can be calculated relatively easily, can be adapted for polynary mixtures. For methanol distillation it will be sufficient to consider the interaction between the main components and between the trace components and the two main components. The interaction between the trace components is negligible and can be disregarded for plant design purposes under the aspect of product and effluent purities.

As is well known, the stripping factor in the distillation section of a column is that figure which indicates whether a certain component tends to migrate towards the column top or as a liquid into the column bottom. The stripping factor is defined by

$$S = V/L \cdot K,$$

i.e. it is determined by the vapour/liquid ratio at a certain point along the column and by the equilibrium constant K. If the stripping factor is greater than 1, the respective component tends to concentrate in the overhead vapours, if it is smaller than 1, it tends to concentrate in the liquid flowing downward. This means that the equilibrium constant K must be greater than the ratio L/V if a certain component is to be distilled overhead. Figure 4.4 shows an L/V ratio of 3.2 at the raw methanol inlet to the prerun column. This line intersects the equilibrium curve for C_9 at a K value of about 3.2, and the value on the abscissa shows that this corresponds to a molar water percentage of 0.23, i.e. some 14.4 wt. %, in the methanol/water mixture. Hence, the intersection corresponds to a stripping factor $S = 1$.

In order to force the n-decan into the vapour phase, the water percentage would have to be 0.38 mol/mol, i.e. the methanol/water mixture would have to be enriched to approximately 25.6 wt. % of water. As raw methanol from coal normally contains between 5.0 and 12.0 wt. % of water, the maximum process water rate to be recycled from the bottoms of the pure methanol column to the prerun column would have to be 5 times the original water content in the raw methanol if the n-decane were to be reliably removed from the raw methanol already in the prerun column. In addition to the extra expenditure on equipment and energy that would be required in such an arrangement, this multiplication of the water content would also produce another adverse effect. As shown by Fig. 4.4, the volatility of ethanol would also increase as the water content of the methanol increases and it would thus become more difficult to remove the ethanol at the pure methanol distillation stage. As the higher hydrocarbon content in the raw methanol is in the range of only a few ppm, hydrocarbons of C_{10} or more are allowed to break through into the pure methanol column where they can be removed through appropriately arranged side outlets with negligible losses of methanol.

The second key component in this system is methyl formiate. Acetone occurs in the raw methanol only in the order of a few ppm, and dimethyl ether is much more volatile than methyl formiate and can therefore be separated more easily from the raw methanol. Essentially, only the prerun column is a pure stripping column, the heat supply to the column bottoms and the number of trays being designed to ensure that methyl formiate is, for all practical purposes, removed completely from the bottom product, the so-called stabilized methanol. Refluxing the methanol condensed in the overhead condensers is inevitable to keep overhead methanol losses to a minimum.

Heating the prerun column with its approximately 40 trays requires some 0.86 GJ per ton of methanol or about 390 kg of low-pressure steam. This results in a reflux ratio to the cold feed of about 1 : 3 at the column head. The low-pressure steam of at least 2 bar may be replaced also by waste heat, for instance from the raw gas, at temperatures above 120°C.

The bottoms product from the prerun column is fed between tray 68 and 72 of the pressurized pure methanol column which contains a total of 80–85 trays. This column operates at a pressure of 7–8 bar corresponding to a bottoms temperature of 125 to 135°C, depending on the water content of the raw methanol.

In order to keep the reboilers down to a reasonable size, the column has to be heated either with steam at a pressure of not less than 3.5 to 5 bar or with waste heat at a temperature level above 140–150°C. Unlike the prerun column, the pressurized column is a genuine distillation column as the overhead product has to meet the purity requirements of US Grade AA methanol. The reflux rate, the number of trays and the heat input can be varied within certain limits, and the most favourable design of the column and its economical operation have to be established by optimizing calculations. A column with the above-mentioned number of trays reaches its operating optimum with a reflux ratio of approximately 3.0 and a heat input of about 2.0 GJ per ton of total methanol produced. As the overhead product from the pressurized column is used to heat the atmospheric column, either of the two columns has to be used to distill some 50 % of the total methanol produced, except for slight differences in the reflux ratio.

In view of the fact that the water content in the bottoms product of the pressurized column is twice that of the prerun column, the hydrocarbons transferred from the prerun column into the pressurized column will reliably be found in the bottoms product, i.e. they are transferred to the atmospheric distillation column. Thus, ethanol becomes the key component for the pressurized column. Since the bottom product of the pressurized column – unlike that of the atmospheric column – does not have to meet certain purity requirements, this column need not have a side outlet for ethanol, but the ethanol is quantitatively transferred to the atmospheric column. The high methanol content in the bottom of the pressurized column facilitates ethanol separation. Nevertheless, for the same number of trays, the pressurized operation of this column leads to a higher reflux than in the atmospheric column. The bottoms product from the pressurized column is transferred to the atmospheric column at approximately 125–35°C. The overhead product is obtained at approximately 115–125°C, condensed in the reboiler of the atmospheric column, and fed to the reflux drum of the pressurized column. From there, some of the overhead product is withdrawn by way of an after-cooler as "on-spec" methanol while the rest is pumped back uncooled as reflux to the column head.

The feed is delivered to the atmospheric distillation column between trays 58 and 62; the column has a total of 80–85 valve trays. The operation of this column is governed by the requirement that the overhead product should not contain more than 10 ppm of ethanol and 1–3 ppm of hydrocarbons while the water collected in the bottom should be largely alcohol-free. This design basis proved to be more cost-effective than the downstream addition of a so-called water purification column for the bottoms product of the atmospheric column as it has some times been practised. In order to meet the purity requirements for the overhead and bottom products, ethanol and hydrocarbons have to be withdrawn through side outlets. This can be achieved best, i.e. with minimum losses of methanol, if the two components are withdrawn at the points where their concentrations are highest. Figure 4.5 shows the concentration profiles of methanol, water, ethanol and n-decane for the theoretical number of trays required by the operating conditions for the example as described here. In this system, the atmo-

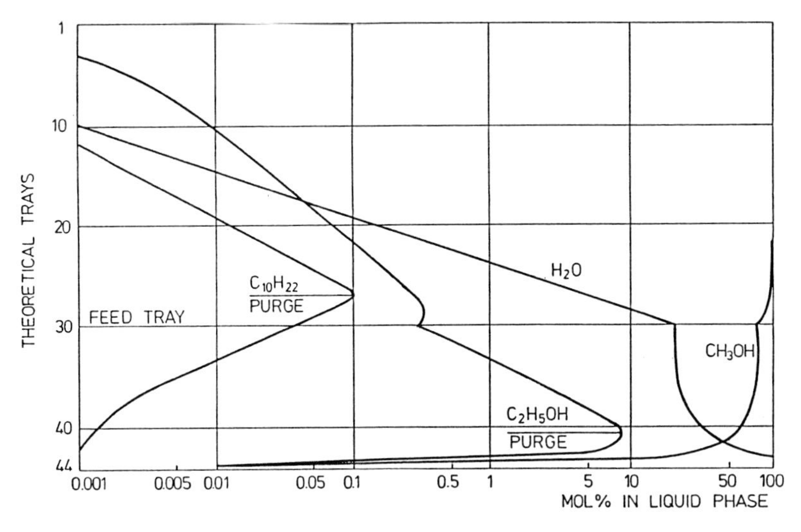

Fig. 4.5. Component profile of a refining column using ethanol and *n*-decane purge

spheric column operates at a pressure of about 1.6 bar at the bottom and a cold reflux ratio of 3.0.

The raw methanol considered in Fig. 4.5 contained 11.0 wt. % of water, 0.08 wt. % of ethanol and 10 ppm of *n*-decane. These figures were roughly doubled in the feed to the atmospheric column as some 50 % of the methanol content of the raw methanol had been boiled off in the pressurized column. As shown by the diagram, ethanol is withdrawn from the 41st theoretical tray, i.e. below the feed tray, while the hydrocarbon is taken off the 27th tray, i.e. above the feed tray. These are the points where the concentrations of the two substances clearly reach their peaks. The quantity of methanol lost with the C_{10} is negligible; it amounts to less than 0.1 wt. %. Some 0.42 wt. % of methanol go to the fuel system together with the ethanol.

The alcohol content in the waste water is less than 100 ppm, the pure methanol does not contain any hydrocarbons, the ethanol content is clearly less than 10 ppm and the water content is about 50 ppm.

The operation of the atmospheric column is more or less the same as that of the pressurized column. As described above, the column is normally heated exclusively by condensing the overhead vapours from the pressurized column in appropriately designed reboilers. Usually the system includes a trim reboiler heated with low-pressure steam to make up for any imbalances occurring during operation. The overhead vapours are normally condensed in air condensers and the methanol is further cooled down with water.

A number of suggestions have been made by which the economics of methanol distillation might be further improved. For example, the overhead vapours from the prerun column, too, might be used for heating the atmospheric column. Although this would reduce the overall distillation heat demand from 2.86 GJ corresponding to 1.3 t of low-pressure steam to 2.0 GJ per ton of pure methanol, it would mean that heat would have to be supplied to the prerun column at a higher temperature level. In addition, the plant would lose some flexibility because of this additional loop. Another suggestion is to use a heat pump (vapour compressor) in order to heat up the overhead vapours from the atmospheric column, which are condensed at approximately 70°C, to a higher temperature level and use them again for heating purposes or for heating an absorption chiller unit using for instance lithium bromide as a refrigerant. The syngas could thus be cooled down below ambient temperature to save energy at the syngas compressor, or the chilling capacity could be used for other applications. In addition to their higher operating expenditure, such systems suffer normally from the disadvantage that the capital investment required to install them cannot be recovered by corresponding savings in production costs.

ICI has proposed their *Higee Distillation System* which is said to substitute a distillation column at lower capital cost. No commercial apllication has been reported so far [4.1].

4.4 Purification of Mixed Alcohols

As mentioned already in the previous chapter, an alcohol mixture which is to be used as octane booster for motor fuels must not contain more than approximately 0.5 wt. % of water. It must also be "stable", i.e. storable. Alcohol/water mixes, which contain much higher quantities of water, must therefore be purified to remove the low boilers and distilled to reduce their water content.

For their *MAS* product, SNAM suggest purification in several stages [4.2]. As shown in Fig. 4.6, a first distillation stage is used to separate methanol and all low boiling components from the high boilers. At this stage, methanol is withdrawn from the distillation column separately from the low boilers and un-

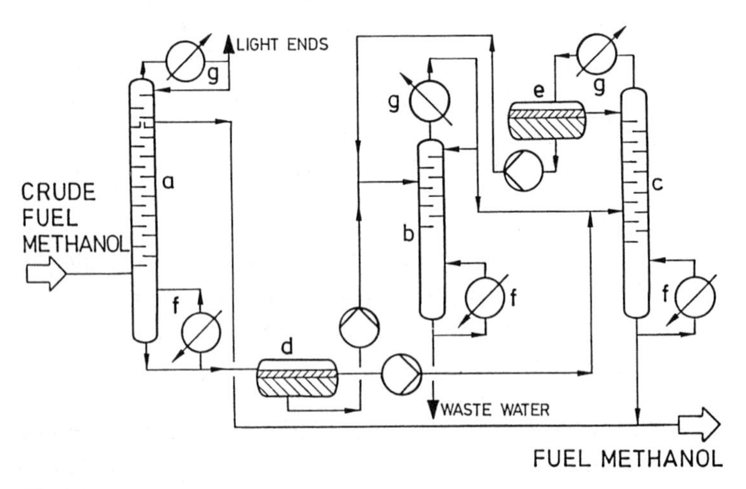

Fig. 4.6. Purification of fuel methanol. (*a*/*b*) Distillation columns; (*c*) azeotropic distillation column; (*d*/*e*) phase separators; (*f*) reboiler; (*g*) condenser

condensible components. The high boilers, i.e. water and higher alcohols are collected in the column bottoms.

The bottoms product is cooled and fed to a first phase separator to form a light organic and a heavier aqueous phase. This aqueous phase, which still contains some organic components, is transferred to a second column in which the organic phase boils off overhead while the water is withdrawn from the bottom. The distillate is mixed with the organic phase from the first phase separator and fed to a third column for azeotropic distillation with cyclohexane, producing an organic phase with low water contents at the bottom. This organic phase is mixed with the methanol from the first column to obtain the almost water-free MAS product.

The overhead product of the third column is an azeotropic mix of higher alcohols, cyclohexane and water which is fed to a second phase separator. The light phase in this separator, consisting essentially of cyclohexane, is recycled to the second column whereas the aqueous phase is fed either to the first column or to the first phase separator. The diagram does not show the equipment to handle the cyclohexane and to remove the residual alcohol from the waste water. This plant arrangement yields a product containing less than 0.1 wt. % of water.

As the raw product from LURGI's *Octamix Synthesis* does not contain more water than can be tolerated in a blending component to natural motor fuels, a marketable product can be obtained simply by eliminating the uncondensibles and those components with boiling points lower than methanol, insofar as they may cause trouble during transportation or storage. This is done in a low boilers column similar to that for pure methanol distillation [4.3].

If, on the other hand, it is desirable to obtain a high concentration of higher alcohols in the end product, or if pure methanol is to be produced along with fuel methanol, a combined low boilers/methanol column is used. With this arrangement, methanol and low boilers boil off from the bottom section, leaving an alcohol mix whose content of high alcohols can be adjusted almost arbitrarily. The methanol is withdrawn through a side outlet at the upper end of the column bottom section, while the low boilers are expelled overhead in the usual way through the top section of the combined column. The methanol, which contains some low boilers, may then conveniently be purified to Grade AA methanol. It should be noted, however, that this operating system also increases the water content of the bottoms product in proportion to the alcohol enrichment.

If this water content in the bottoms product exceeds the maximum that can be tolerated for motor fuel components, it may be corrected by using a conventional molecular sieve. Such molecular sieves can reduce the water contents to as little as 200 ppm.

5. How to Process By-Products and Wastes

More and more stringent clean water and clean air legislation has also changed the general attitude with respect to byproducts and pollutants from industrial premises. Whereas until less than two decades ago, attention was focussed mainly on the most cost-effective production of marketable byproducts such as sulfur, tar and oil products, phenols and ammonia, priority today is not under all circumstances on the cost-effectiveness of the respective processes. Owing to rigorous limits on pollutant contents in waste gases and waste water, the residual by-product and pollutant contents in the effluent streams play at least as big a role as the percentage to which valuable products can be recovered.

This chapter will therefore deal not only with recovery systems, but in somewhat greater detail also with processes used to clean the waste gases and waste waters.

5.1 Water and Aqueous Condensates

For the purposes of this chapter, the word *waste water* will be used for all effluents discharged from the various sections of coal-based methanol plants whose main component is water. They include aqueous condensates from direct condensation of the water contained in the raw gas, residual steam which is first removed from the gas by appropriate absorbents and then desorbed again by distillation, as well as the water produced in the chemical reactions, such as the process water resulting from the reaction of carbon dioxide with hydrogen in the methanol synthesis loop, which is removed by the methanol distillation stage. Waste waters in a broader sense include also the blowdown water from the steam boilers and the water used to regenerate the ion exchangers in the B.F.W. conditioning units. As the latter two do not contain any substances which may lead to conflicts with environmental legislation, they can be added without further treatment to the normally much larger waste water stream from the plant as a whole.

Wherever it is economically justifiable, the waste waters are largely reused within the plant so that as little as possible has to be discharged beyond the battery limits. Reusing may take the following form:

- suspension agent for coal breeze
- quenching water for direct raw gas cooling

Table 5.1. Typical waste water specification Fed. Rep. of Germany 1983

Temperature	max.	30°C	
pH-Value		5-9	
Solids	max.	100	mg/l
BOD 5	max.	100	mg O_2/l
COD	max.	500	mg O_2/l
H_2S	max.	1	mg/l
HCN	max.	0.5	mg/l
CNS^-	max.	5	mg/l
NH_3 (as N)	max.	100	mg/l
Phenols (total)	max.	20	mg/l
Mercaptanes	absent		
Carbon	absent		

- treatment to serve as cooler makeup water
- treatment to serve as boiler feed water.

The higher the required quality of the water, the higher will be the treatment cost, above all for process condensates containing substances which are difficult to eliminate. The water to be discharged has to meet authority requirements which may be quite different from country to country. No general specification can therefore be given. Instead, Table 5.1 illustrates a typical waste water specification as it was valid in the Federal Republic of Germany since 1985.

5.1.1 Treatment of Aqueous Gas Condensates

Whatever the processes to produce methanol syngas from coal may be, the hot raw gas always contains steam, irrespective of the raw material and method of gasification. This raw gas steam content comes from the moisture of the coal, from unreacted gasifying steam and from water formed by oxidation of the hydrogen contained in the coal during the gasification process. Aqueous condensates are obtained when the raw gas is cooled. These condensates may be more or less polluted, depending on the type of coal and gasification process used. Table 5.2 provides typical data for two different raw materials and different processes – medium-temperature, countercurrent gasification of lignite, and high-temperature gasification of a bituminous coal/water slurry. It can be seen that the condensate quantity and pollutant contents increase as the gasification temperature decreases.

Mechanical treatment is normally the first step in the waste water treatment process. Suspended solids (coal breeze, ash, slag) and emulsified substances (tars, oils) whose specific density differs from that of water are eliminated from the aqueous condensates by mechanical means. Condensates from medium-temperature gasification undergo a first rough treatment in tar separators – mostly simply atmospheric vessels which are normally equipped with devices to ensure an even flow distribution and to facilitate phase separation, and with separate draw-off systems for tars, oils and water. Occasionally, clean tar and

Table 5.2. Gas condensate – typical data

		Lignite Medium Temperature Gasification	Coal Slurry High Temperature Gasification
Quantity	(t/t of coal)	0.7 - 1.2	0.6 - 0.7
Suspended matter	(mg/l)	1000	600
Sulfur	(mg/l)	600	100
Chloride	(mg/l)	50	500
NH_3- + NH_4-Ions	(mg/l)	10,000	400
Cyanide	(mg/l)	50	5
Phenols	(mg/l)	4,000–5,000	–
COD	(mg O_2/l)	10,000	60

dust-laden tar are separated in the same vessel. Figure 5.1 shows one of several possible tar separator designs.

The condensate is fed to a central downpipe and evenly distributed by appropriate internals over a hood-type baffle. The tar runs down over this baffle to be collected in the bottom of the vessel. The dust-laden tar is then withdrawn by the central bottom nozzle, while the lighter clean tar rises to the surface and flows off through the clean tar nozzle. The oil, having a lower specific density than water, forms the supernatant phase and is collected in the upper hood. Like the clean tar, it rises to the surface and is withdrawn through the oil outlet.The water leaves the tar separator about half way up between the oil outlet and clean tar outlet. The dust-laden tar settling at the bottom may be recycled to the coal gasification section.

If the aqueous effluent, which may still contain as much as 100 mg/l of solids and up to 5 000 mg/l of suspended oils and tars at the tar separator outlet, is to be cleaned even further, vessels having parallel plate packs installed can be installed with good results. These parallel plates produce a laminar flow in which particles exceeding 0.01 mm in size can settle. A typical example of such a laminar pack separator is the *CPI separator* developed by SHELL [5.1] In this separator, oily water flows through a corrugated plate pack installed in a concrete structure at an angle of approximately 45°. The water enters the separator at the lowest point of the plate pack, the oil is coagulated at the bottom of the plates and rises to the surface between them so that it can be skimmed off. Solids are quickly sedimented by the laminar flow within the plate pack; they slide downward into a crock sump from which they are removed at regular intervals. This system ensures that the residual content of oils and sediments can be reduced to less than 20 ppm.

Wherever such a precleaned waste water is to be further treated, for instance by extraction or distillation, the solids content has to be reduced to a much lower level. This is done not only in the pressurized gravel bed filters which have been standard equipment for decades already although they tolerate only small

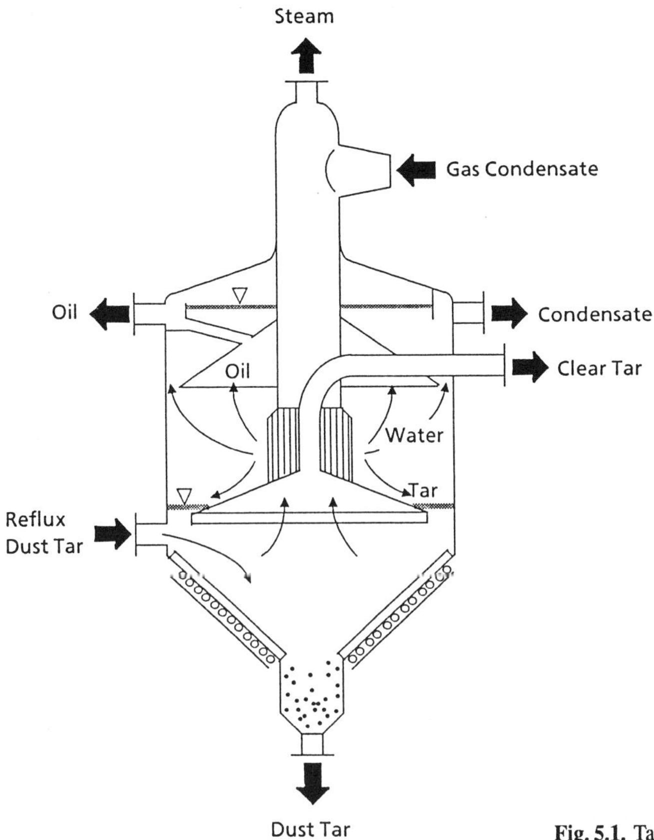

Fig. 5.1. Tar separator

specific loads, but also in modern types of pressure leaf filters, pressure candle filters and pressure rotary drum precoat filters which can achieve residual solids contents of less than 5 ppm with very high specific loads. These filters also make the residual tars and oils coagulate so that they are backwashed from the filter surface together with the solids. Whenever the aqueous gas condensates contain high loads of dissolved organic substances, the phenols have to be eliminated in order to reduce the oxygen demand (BOD) in the biological treatment stage. Either extraction or adsorption stages can be used to this end.

The LURGI *Phenosolvan Process* by its very nature is an extraction process to recover raw phenol which can then be broken down in other units into its individual components such as phenols, cresols, pyridine bases and many more as required under the prevailing market conditions. Whereas other phenol removal processes used in the past could eliminate only the volatile (mono-valent) phenols from the waste water, the Phenosolvan process is suitable also to extract most of the multi-valent phenols. The extraction agents are butyl acetate or diisopropyl ether. Although the phenol distribution factor is much lower for the latter than for butyl acetate, its lower boiling point permits regeneration at a lower temperature. Regeneration also requires less steam. Propyl ethers are much more stable

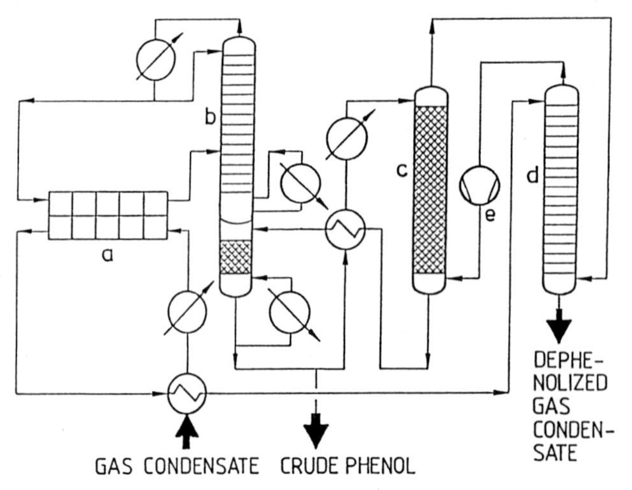

Fig. 5.2. *LURGI* phenosolvan process. (*a*) Mixer/settler extractor; (*b*) distillation; (*c*) scrubber; (*d*) solvent stripper; (*e*) blower

chemically than butyl acetate so that solvent losses caused by hydrolysis, which are about 150 g/m^3 of waste water for butyl acetate, are not more than about 20 g/m^3 for propyl ether.

Figure 5.2 shows the operating principle of a Phenosolvan unit. The condensate, which is largely free from tar and oil, is fed to a multi-stage mixer/settler battery to be dephenolated in a countercurrent with diisopropyl ether. Mixing is normally ensured by submerged pumps accommodated in a so-called mixing chamber; the settling chamber consists of a cylindrical or rectangular space arranged between two mixing chambers. The operating principle of this setup may be explained by looking at Fig. 5.3. The heavier of the two liquids enters the pump pit of the first mixing stage and is mixed there with the lighter liquid overflowing from the settling chamber of the second stage. The lighter phase overflows into a funnel and is discharged from the facility. The heavier phase flows through the subsequent stages in a countercurrent to the lighter phase fed to the mixing chamber of the last stage. It finally leaves the settling chamber of the last stage as raffinate.

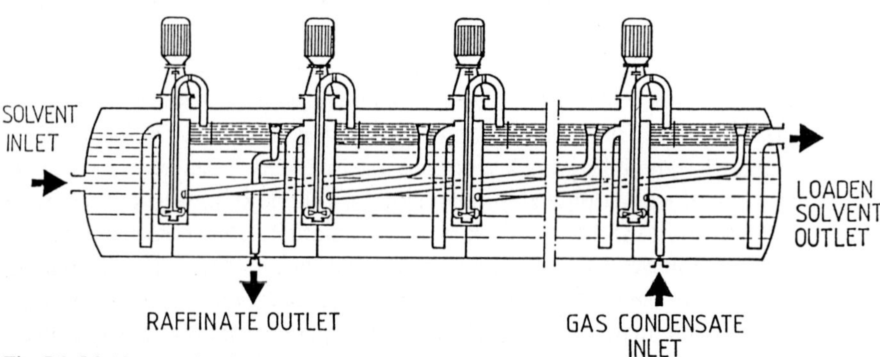

Fig. 5.3. Multistage mixer/settler extractor

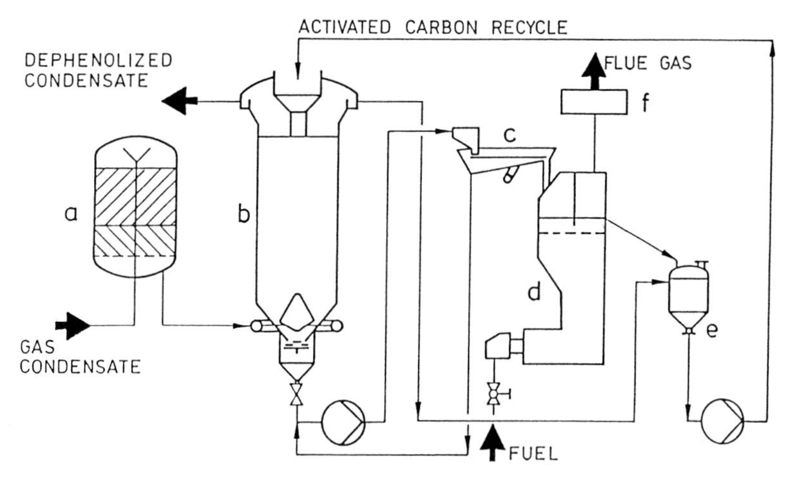

Fig. 5.4. *Bergbauforschung* process for dephenolization by activated carbon. (*a*) Filter; (*b*) adsorber; (*c*) sieve; (*d*) fluidized bed regenerator; (*e*) quench; (*f*) combustor

The dephenolated raffinate is then fed to the top of the solvent stripper to recover the solvent by stripping with gas. To remove the solvent from the stripper overhead gas, the latter is washed with raw phenol, and the solvent-free gas is recycled to the stripper. The extract from the mixer/settler, containing the phenols and other extracted components, is fed to a two-stage distillation unit to produce highly pure ether overhead and solvent-free raw phenol at the bottom. Depending on the bottom temperature of the distillation column, the raw phenol may contain up to 5 % water. The raw phenol is partly exported as a product and partly used as wash liquor for the solvent scrubber. The ether-laden bottom product from the solvent scrubber is recycled to the lower stage of the distillation column to expel the ether.

Whereas the Phenosolvan process primarily aims at recovering the phenols from the waste water, the *Activated Carbon Process* developed by *Bergbau-Forschung* destroys these phenols. As illustrated in Fig. 5.4, the gas liquor first passes a two-stage filter before it is fed to the bottom of an activated carbon adsorber. As it rises through this adsorber, the activated carbon – which may in certain cases be replaced by coke – adsorbs the phenols from the water so that the dephenolated gas liquor leaves the adsorber at the top. The activated carbon is regenerated continuously. A mixture of gas liquor and activated carbon is withdrawn by way of a rotary valve in the adsorber bottom and pumped onto a screen with oblong mesh where the gas liquor is separated from the activated carbon to be recycled to the pump loop.

The activated carbon retained by the screen is fed to a fluid bed regenerator in which hot flue gases are blown through it. The phenols and all other substances which are volatile at the prevailing temperature are thus desorbed and discharged from the regeneration system. These substances are then burnt in an incinerator. The regenerated activated carbon overflows at the top of the fluid bed, is quenched with dephenolated gas liquor and pumped back to the adsorber.

Tests aiming at recovering the phenols by carbon extraction have so far been unsuccessful. It is generally possible, however, by treating the condensate with activated carbon, to bring down the COD to a lower level than this would be possible by liquid-liquid extraction. Economic factors limit the application of this process to relatively small waste water quantities with high phenol concentrations.

5.1.2 Removal and Recovery of Sulfur and Ammonia

The volatile components of aqueous condensates are for the most part CO_2, NH_3 and H_2S. HCN is normally present as well. Some of these substances occur in complex compounds such as NH_4OH, $(NH_4)_2CO_3$, $(NH_4)CO_3$, $(NH_4)_2S$, NH_4HS or NH_4CN.

Normal practice in the past was to expel all volatiles together in a total stripper. The vapours were then washed with sulfuric acid to produce ammonium sulfate. However, as the latter is more and more difficult to sell as a fertilizer, the NH_3 is now recovered separately from the sour gases in pure form. This separation can basically be achieved by two methods:

- In a purely physical process based on the different evaporation behavior of NH_3, H_2S and CO_2 in aqueous solutions at different pressures and temperatures;
- in a chemical process expelling the NH_3 from the total stripper vapours by means of a solvent which releases the NH_3 again under different conditions and does not absorb the sour components.

A very typical example of a purely physical process is the *CLL-Process* developed jointly by *Chemie Linz AG* and LURGI. Figure 5.5 illustrates a simplified flow diagram of this process, showing that the gas condensate is split up into 4 streams:

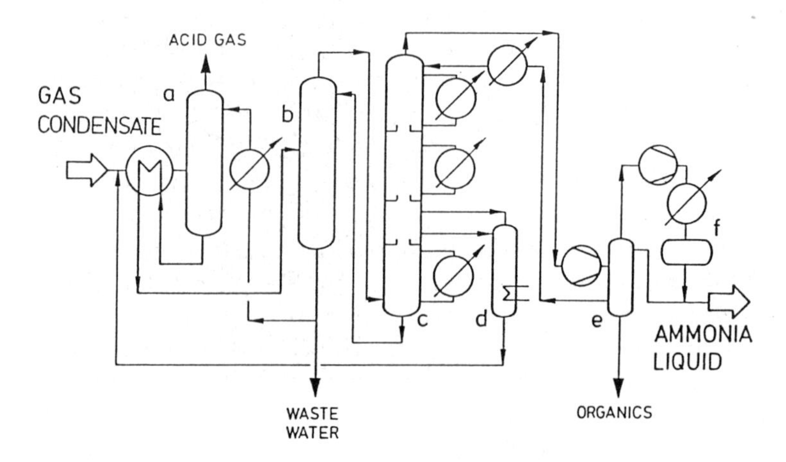

Fig. 5.5. CLL process. (*a*) Deacidifier; (*b*) total stripper; (*c*) CO_2 scrubber; (*d*) NH_3 stripper; (*e*) dryer; (*f*) NH_3 condenser

- stripped gas condensate containing less than 100 ppm free NH_3 so that it can be discharged directly to the biological treatment section
- pure liquid ammonia
- an acid gas consisting of CO_2 and H_2S, which is virtually free from ammonia and can be fed to a *Claus* unit
- low-boiling organic impurities (alcohols, ethers, etc.) occurring in small quantities which can be incinerated.

The process diagram shown here is used for such gas condensates whose sour gas content is as least as high as their ammonia content. The $NH_3/CO_2/H_2O$ system exhibits different boiling properties at different pressures. The so-called maximum temperature azeotrope lines, which are defined so that the CO_2/NH_3 ratios are the same in both phases when the liquid and gas phases are in equilibrium, are shifted in the direction of higher ammonia mixes as pressures are increased.

The dephenolated gas liquor is fed to the deacidifier whose pressure is appropriately set for the CO_2/NH_3 ratio of the gas liquor. The acid gases are stripped with steam and then washed with water before leaving the deacidifier in order to remove the remaining NH_3. The sour gas is obtained at a slightly elevated pressure and contains less than 100 ppm NH_3. It can be fed straight to a *Claus* unit or to a wet catalysis unit to produce sulfur or sulfuric acid.

The deacidifier bottom product, which still contains a small quantity of sour gases, is fed to the top of the total stripper to expel all the gaseous components which are still in the condensate. The condensate leaving the total stripper bottom is cooled and can then be discharged to a biological treatment unit. It contains less than 50 ppm of NH_3 and less than 1 ppm of H_2S. Its BOD is less than 20 ppm. Low-salt condensates as they are normally obtained from coal gasification may be used as makeup water for cooling water systems.

The vapours withdrawn at the total stripper head are partially condensed in the acid scrubber and the condensate is then recycled to the total stripper head. The uncondensed vapours are fed to the acid scrubber where the sour components are removed in several condensation and wash stages by means of aqueous ammonia fed to the scrubber head. In addition to ammonium carbonate and ammonium sulfide, the acid scrubber bottom product contains a surplus of free ammonia which is thermally expelled in the ammonia stripper. The quantities of CO_2 and H_2S recycled between the deacidifier and the total stripper are thus substantially reduced and the heat economy of the system is improved. The ammonia stripper bottom product is fed to the deacidifier to remove CO_2 and H_2S. The acid scrubber overhead product consists of ammonia containing less than 4 wt. % of water, less than 0.1 wt. % of carbon dioxide and less than 1 ppm of H_2S. This stream also contains all low-boiling organic substances.

The *Phosam-W Process* developed by *US-Steel* is essentially a chemical process although – as illustrated by Fig. 5.6 – it is preceded by a physical stage. The bottom of what has been termed a *superstill* is designed as a stripper section in which the sour gases and the ammonia are jointly expelled from the aqueous condensate which is then delivered from the stripper bottom to the biological

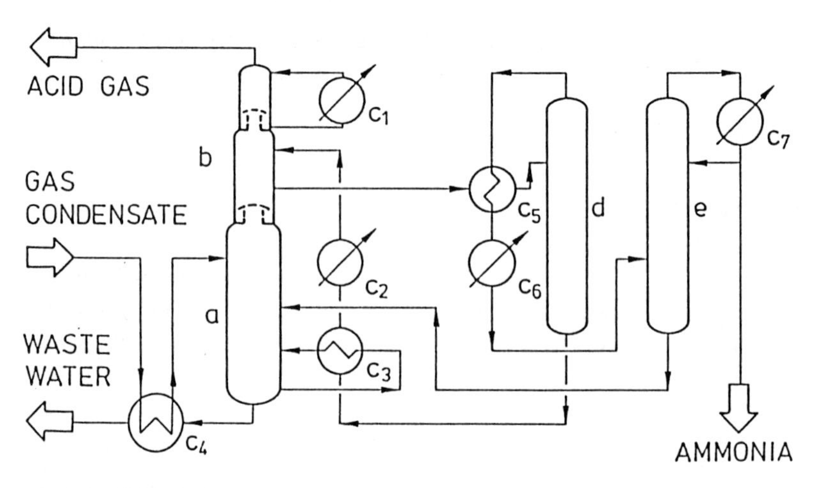

Fig. 5.6. PHOSAM-W-process. (*a*) Sour water stripper; (*b*) absorber; (*c*) 1–7 condenser/heat exchanger; (*d*) stripper; (*e*) fractionator

treatment unit.The ammonia is then selectively removed by an aqueous solution of ammonium phosphate – termed Phosam solution – from the gas/vapour mix in the absorber top of the superstill. This selective absorption is achieved by a reversible chemical reaction between ammonia and the ammonium phosphate solution whose composition changes from mono- to diammonium phosphate. Steam is used to expel the ammonia from the spent Phosam solution in the stripper. The ammonia is fed to a fractionator while the regenerated solution is pumped back to the absorber section of the superstill.

The vapours from the absorber top contain large quantities of steam and up to 0.4 % of ammonia. If their sulfur content is so high that the vapours cannot be incinerated either alone or together with other gases but have to be treated in a sulfur recovery unit, for instance by the *Claus* process, the absorber top is followed by a condensation stage. The ammonia content of the vapours is thus reduced to 1 000 ppm so that the H_2S/CO_2 mix, from which most of the steam has been eliminated as well, can be conveniently treated in a *Claus* unit. The vapour condensate contains large quantities of NH_3, almost all the HCN that has entered the plant with the raw gas, as well as the organic components of the gas liquor. Fractionated distillation in the steam-heated fractionator concentrates the aqueous ammonia to 99.99 wt. % NH_3. In order to obtain such highly pure NH_3, the other acid gas components are causticized in the fractionator with caustic soda. The ammonia so obtained meets the requirements for a synthetic product. The treated gas liquor leaving the total stripper contains more or less the same quantity of impurities as the CLL process described in the previous chapter.

Particularly attractive for precleaning of condensates containing phenols, ammonia and sour gases is a combination of the Phenosolvan and CLL processes. Neither the solvent stripper nor the gas loop are required. The solvent is recovered in the deacidifier. The ether contained in the sour gas leaving the deacidifier overhead is removed by means of raw phenol and recycled to the extraction stage by way of the bottom section of the solvent distillation column.

5.1.3 Biological Treatment

The aqueous condensates normally require a secondary treatment with microorganisms to clarify them sufficiently for discharge to public water courses. These microorganisms use oxygen to break up the dissolved and colloidal organic impurities, their metabolism producing endogenic substances termed activated sludge, and H_2O plus CO_2 as waste products. Depending on the operating conditions, the impurities may be converted even to minerals.

Among the two processes known

- the *activated sludge process*
- and the *trickling filter process* [5.1],

the latter is rarely used because it requires a very uniform quality of appropriately pretreated water to function properly.

The activated sludge process uses an enriched bacteria suspension to which normal or O_2-enriched air is added. Surface aerators blow the oxygen required for microbiological decomposition into the water of the activated sludge tank. A downstream settling tank is used to separate and prethicken the bacteria suspension. The following criteria are essential for smooth operation of the facility

- constant pollutant concentration, salt content and temperature
- no sour effluent surges
- no intake of toxic substances
- sufficient oxygen and nutrient supply.

An upstream buffer or equalization tank is used to even out variations in pollutant concentrations. The biological process is jeopardized not by the absolute salt concentration but rather by excessive fluctuations in the salt content which may destroy the flocculated bacteria suspension.

Sour effluents need to be *neutralized* under all circumstances before they are allowed to enter the activated sludge tank, whereas an alkaline waste water inrush rarely disturbs the system because the steady production of CO_2 as a consequence of microbic activity tends to neutralize the system automatically. *Toxic substances* have to be eliminated by a specific pretreatment. Flocculation and adsorption processes have been found suitable for this purpose. Under special circumstances, the respective effluent stream may even have to be branched off for separate treatment.

Appropriate measuring and control instruments have to ensure a sufficient oxygen supply, and almost all effluents require additional doses of nutrients – mostly phosphorus but occasionally also nitrogen compounds. The nutrient balance is normally supported to a certain extent by sanitary sewage, although the quantities are relatively small. The treatment tanks are normally arranged horizontally, but vertical designs have also been developed and used in individual

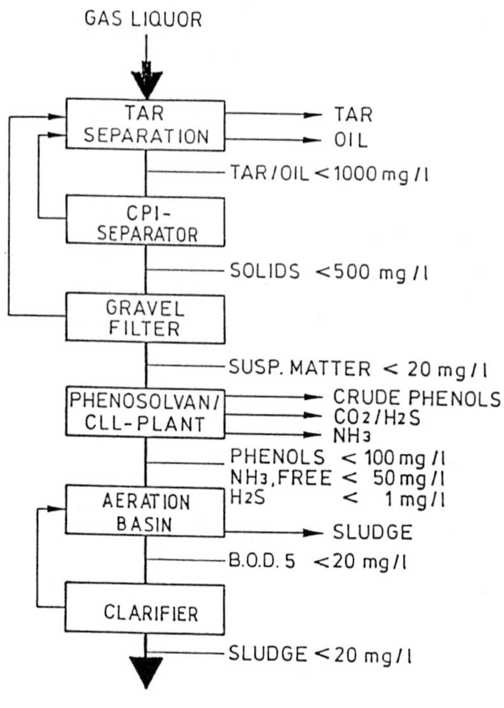

Fig. 5.7. System of gas liquor treatment

cases. Their shaft depths of as much as 100 m are not only space saving but offer the additional advantage that they are largely inodorous.

A number of additional process steps can be applied to specific water treatment problems

- multi-media filtration
- adsorption by activated carbon
- membrane filter processes.

A skilful arrangement may produce waste water qualities that meet even stringent requirements, for instance as boiler feed water, and it is theoretically possible to realize a *zero effluent* system which produces only very small quantities of regenerates.

A typical example to describe the possible combinations of waste water treatment is shown in Fig. 5.7 illustrating the treatment of gas liquor from medium-temperature pressure gasification of bituminous coal. The aqueous effluent is to be used as makeup water for an open cooling loop so that no conflicts with environmental legislation can be tolerated either from evaporation or from blowdown.

Upstream of the tar separator, the gas liquor is composed approximately as follows:

- tar and oil 15.0 g/l
- Solids 2.0 g/l
- phenols, mono- and multivalent 4.0 g/l
- low-boiling organic components 0.3 g/l
- NH_3 15.0 g/l
- CO_2 35.0 g/l
- H_2S 0.5 g/l

Possible byproducts are tar, oil, raw phenol, ammonia and sulfur.

5.2 Hydrocarbon Condensates

Countercurrent gasification of coal produces not only high-boiling tars and oils together with the condensed residual steam, but also low-boiling products which can be eliminated from the raw gas only by cooling it below ambient temperature or by a scrubber. To remove these low-boiling hydrocarbons, which are normally termed *gas naphtha*, prewashing with cold methanol as described in Chap. 2 has been found effective.

The condensates from the various gas cooling stages and from the prewash section are not products with clearly defined boiling ranges. They contain dust and water which have to be eliminated before the condensates can be further processed. The solids are removed by conventional means; the water – unless it can be eliminated by phase separation – has to be removed by distillation. To this end, a distillation column known as a *drier* precedes the hydrocarbon distillation stage. In order to obtain a maximum quantity of valuable naphtha, the two low-boiling fractions from this distillation stage, i.e. light and heavy tar naphtha, are hydrogenated together with the gas naphtha from the methanol prewash stage. The tar fraction boiling above 210°C may be used as fuel oil or to impregnate wood. Catalytic hydrogenating refining saturates olefins and diolefins to paraffins, hydrogenates organic sulfur components, such as mercaptanes and thiophenes, and reacts the nitrogen content of pyridines and amines to ammonia.

A catalytic refining process that has proved its suitability is the *Benzoraffin Process* developed jointly by BASF, LURGI and VEBA. This process hydrogenates gaseous naphtha in the presence of hydrogen at a pressure of approximately 50 bar and at temperature around 400°C over stationary cobalt molybdenum or molybdenum catalysts. As shown in Fig. 5.8, preheated naphtha is evaporated into the circulating hydrogen in a mixer. The unevaporated components are recycled to the tar distillation stage. The mixture of circulating hydrogen and naphtha vapours is first passed through a pre-reactor to hydrogenate the easily polymerizing components which may clog the equipment as they are further heated. The reaction temperature at this preliminary stage is about 200–250°C.

The gas is then heated to approximately 350°C by heat exchange with the hydrogenated product before it is fed to a reactor with interim quenching

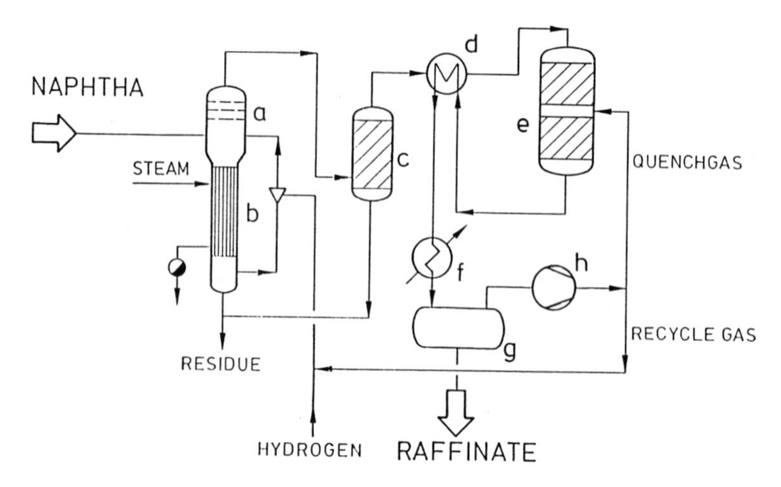

Fig. 5.8. Benzoraffin process for naphtha hydrogenation. (*a*) Evaporator; (*b*) mixer; (*c*) prereactor; (*d*) heat exchanger; (*e*) reactor; (*f*) cooler; (*g*) separator; (*h*) recycle compressor

stage whose first reaction stage contains a cobalt molybdenum catalyst while the second is filled with a molybdenum catalyst to hydrogenate all remaining unsaturated components. A stream of cold recycle gas is used for quenching. The gas flow leaves the hydrogenation reactor at a temperature of approximately 400°C and is then cooled down to ambient temperature in the heat exchanger and in downstream coolers. The raffinate is separated from the circulating hydrogen and the low-boiling components hydrogen sulfide, ammonia, carbon dioxide and water are expelled in a stripper at approximately 10 bar.

The product thereafter contains less than 0.1 wt. % of water and more than 75 wt. % of aromatics. Its sulfur content is less than 200 ppm. The product so obtained is suitable as a blending component for motor gasoline or as a raw material to produce pure aromatics.

5.3 Waste Gases

Like effluent discharge, the emission of waste gases to the environment is strictly limited. The following limits, in particular, have to be observed in the western world for the emission of waste gases:

- dust max. 50 mg/m^3
- special inorganic dusts (e.g. As, Pb, F, Cl, HCN) max. 0.2 - 0.5 mg/m^3
- hydrogen sulfide max. 10 mg/m^3
- ammonia max. 50 mg/m^3
- carbon disulfide, Carbon monoxide max. 100 mg/m^3
- methanol, carbon oxisulfide, paraffinichydrocarbons max. 150 mg/m^3
- sulfur dioxide, sulfur trioxide, nitric oxides max. 500 mg/m^3

However, in addition to these specific limits, the total pollutant discharge, especially of SO_2 and NO_x, is limited as well. For a more detailed discussion, see the relevant statutory regulations [5.3,4].

The major part of gaseous pollutants derived from coal gasification consist of sulfur compounds and among them of hydrogen sulfide. The latter normally accounts for more than 90 % of all sulfur compounds while the rest consists predominantly of COS, mercaptanes and thiophenes. There are normally only traces of CS_2 and SO_2. Exceptions from this rule are high-temperature gasification of coke in a reducing atmosphere which produces about 30 % of the sulfur in the form of CS_2, and flue gases which contain virtually all the sulfur in the form of SO_2. The CS_2 removal from waste gases and SO_2 removal from flue gases will not be dealt with here, as coke gasification with oxygen is not a cost-effective route to produce methanol and as coal-based methanol plants, if used for power generation, usually rely on low-sulfur fuels.

The sulfur components occur in various waste gases; they are all more or less diluted by CO_2. How they are recovered depends on the concentration, and the product is as a rule elemental sulfur. Direct production of sulfuric acid may, however, be interesting in the case of low-sulfur waste gases.

5.3.1 Recovery of Elemental Sulfur

Elemental sulfur can be produced from H_2S-laden waste gases by either of the following two methods:

- oxidizing wet desulfurization, for instance by means of the Stretford Process
- dry, catalytic oxidation according to the *Claus* Process.

The Stretford Process has already been described in Chap. 2. In a way similar to syngas treatment it can be used also to remove H_2S from normally pressureless waste gases. It has already been mentioned that the Stretford Process can absorb virtually no COS. It can therefore be used only for such waste gases whose COS content does not exceed the maximum that can be tolerated for emission to the atmosphere, or else, the H_2S-free waste gas has to be led to an incinerator to convert the COS to SO_2. A process that has proved universally applicable to recover all sulfur components from waste gases is the *Claus Process* in its manifold variations. This process was developed by *C.F. Claus* who filed a patent application for it as early as 1883, originally to incinerate hydrogen sulfide in a single-stage catalytic operation to obtain elemental sulfur according to the equation

$$2H_2S + O_2 \rightleftarrows S_2 + 2H_2O.$$

Later on, the process was subdivided into two stages – both of them catalytic – incinerating H_2S in the first stage followed by an after-treatment under more favourable equilibrium conditions in a second. Today the process encompasses at least three stages.

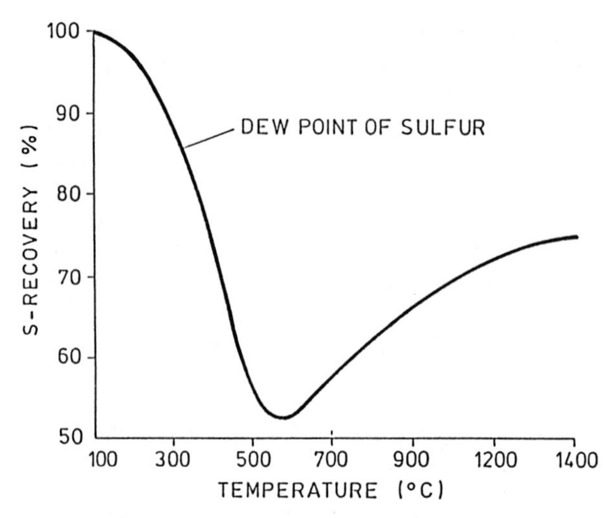

Fig. 5.9. Equilibrium curve for the reaction $2H_2S + O_2 \leftrightarrows S_2 + 2H_2O$ at 1 Bar

The first stage consists of a combustion chamber in which all the hydrogen sulfide is reacted with the stoichiometric air rate to obtain elemental sulfur, sulfur dioxide and residual H_2S. The elemental sulfur is condensed out and the sulfur dioxide and residual H_2S are catalytically reacted in two stages with decreasing temperatures to produce elemental sulfur and water according to the equation

$$2H_2S + SO_2 \rightleftarrows 3/8S_8 + 2H_2O.$$

A number of papers have been published [5.5,6] about the complex equilibria resulting from the different sulfur modifications; these papers show that complete reaction of the sulfur components to elemental sulfur cannot be expected until the temperature is reduced to less than 140°C. Figure 5.9 shows the sulfur yield versus temperature at an overall pressure of 1 bar. It is obvious that the conversion rate drops in the area of purely *thermal incineration* (combustion chamber) and then rises steeply again as the temperatures decrease. As the reaction velocity is very small at temperatures of less than 350°C, catalysts have to be used for low temperatures. Reactions in this temperature range are accelerated by highly active Al_2O_3 catalysts which are normally doped with cobalt and molybdenum to improve CS_2 and COS conversion.

The reaction equilibrium and thus also the achievable sulfur yield are heavily dependent upon simultaneous side reactions caused by such other sour gas components as CO_2, NH_3, HCN and, above all, by hydrocarbons producing COS, CS_2, CO and similar substances. As Fig. 5.9 shows that temperatures of less than 300°C are already close to the dew point of vaporous sulfur, it is not possible to reach a 100 % sulfur yield. The limit for *Claus* units with two catalytic stages is 96 %; it can be increased to something like 98 % by adding a third stage. Newly developed titanium oxide catalysts are expected to increase the sulfur yield to approximately 99 % already in *Claus* units with only two catalytic stages [5.7].

5.3.2 Design Variations of the Claus Process

Claus units have to be tailored to the hydrogen sulfide content of the acid gas and to its ammonia and hydrocarbon contents. Hence, numerous process variations have evolved.

The flow sheet of a *Standard Claus* unit is shown in Fig. 5.10. Essentially, such a unit consists of the combustion chamber section with a waste heat boiler and two reactors. Such plants are normally used for sour gases containing more than some 45 vol. % of H_2S. The sour gas is fed to the combustion chamber together with air or, if required, with fuel gas; it is already in this combustion chamber that as much as 70 % of the hydrogen sulfide are reacted to elemental sulfur. The gas mixture leaving the combustion chamber is cooled down in a waste heat boiler to the inlet temperature required for the first catalytic stage. The temperature is controlled by what has been termed a hot gas bypass.

Sulfur conversion is then continued in the first catalytic stage up to approximately 85 %, reaching temperatures around 300°C. Thereafter, the gas mixture flows through a heat exchanger to a second stage of the waste heat boiler to be cooled down to approximately 170°C. Most of the sulfur is liquefied in this section and withdrawn from the gas flow. The gas is then reheated to the required temperature in a countercurrent to the gas flow leaving the first stage and fed to the second catalytic stage. The inlet temperature may be controlled by an appropriate bypass with a heat exchanger, by hot gas from the combustion chamber (*hot bypass*), or by burning either sour gas or fuel gas. The reaction described above proceeds in the second stage at low temperatures almost to thermodynamic equilibrium. After leaving the second catalytic stage, the gas is indirectly cooled to some 125–135°C either by means of boiler feedwater or in a low-pressure waste heat boiler. When the liquid sulfur has been withdrawn, the gas flows through a special sulfur separator which is normally preceded by an agglomerator to agglomerate the fine sulfur mist. The gas has still to be after-treated in order to meet statutory requirements.

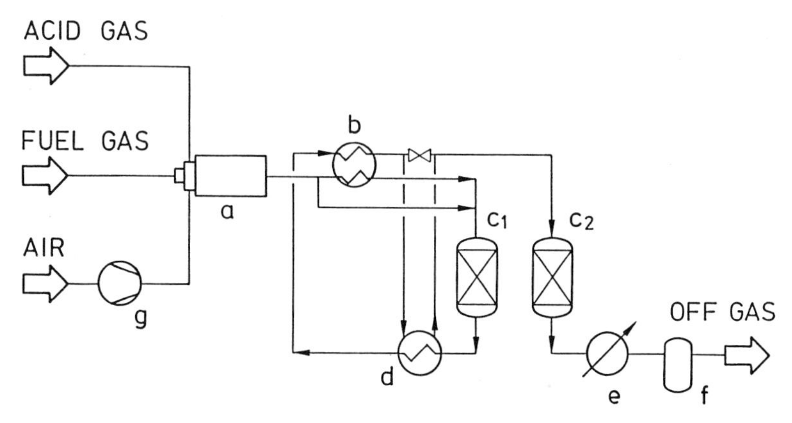

Fig. 5.10. Two stage *standard* Claus plant. (*a*) Combustion chamber; (*b*) waste heat boiler; (*c*) (1/2) reactor; (*d*) heat exchanger; (*e*) sulfur condenser; (*f*) separator; (*g*) blower

The liquid sulfur has to be degassed, i.e. the dissolved H_2S content has to be reduced to figures around 10–20 ppm depending on the locality. As the solubility of H_2S in sulfur increases with rising temperature, it is of great importance that the sulfur should be degassed at as low a temperature as possible. Spraying alone goes far to degas the sulfur. The rest is achieved by, for instance, adding liquid ammonia to the liquid sulfur before spraying (proprietary process of *Société National Elf Aquitaine*). The gases released during degassing have to be extracted and recycled either to the *Claus* unit or to a tailgas treatment unit.

If acid gases containing between approximately 20–50 vol. % of H_2S are to be treated, the Standard Process has to be modified as addition of the stoichiometric air rate would get the concentration of combustible components in the combustion chamber outside the ignition range. Consequently, only a slip stream of the sour gas is fed to the burners in the combustion chamber together with the full air rate, while the rest of the sour gas is fed either directly to the combustion chamber downstream of the burners or immediately upstream of the first catalytic stage.

Whenever the acid gas contains less than 20 vol. % of H_2S, the process has to use either oxygen-enriched air or even pure oxygen. The operating principle is the same as for air, but the burners are different from those in a Standard unit. The use of oxygen also requires more stringent monitoring and safety facilities.

The above concentration limits refer to cold sour gases and cold combustion air. If the gases are preheated, the limits will be shifted downward. The limit for using the *Claus* process is an H_2S concentration around 5 %. Leaner gases can be treated only if the reaction is supported by burning additional sulfur. Nor can gases be treated in *Claus* units if they contain major proportions of hydrocarbons or ammonia. Such gases can only be incinerated and the SO_2 then absorbed or reduced by suitable processes (Sect. 5.3.2).

5.3.3 Production of Sulfuric Acid

As many types of coal contain little sulfur, but have a considerable surplus of carbon for methanol production, the H_2S content of the resulting sour gases is frequently around or even less than 5 vol. %. Although the sulfur concentration can be increased in the gas purification section, this is always a costly undertaking. Certain direct treatment processes may therefore be used for such sulfur gases although they lead to sulfuric acid rather than to the more easily manageable and normally more conveniently marketable elemental sulfur.

The LURGI *Concat Process* offers itself to produce sulfuric acid of up to 93 % concentration from low-sulfur gases. The process tolerates both gases containing the sulfur mainly in the form of H_2S and COS and others which contain almost exclusively SO_2 with only little residual H_2S. As the gas leaving a Concat unit does not contain more than small quantities of SO_2, the process can be used also for tailgas treatment downstream of *Claus* units.

Figure 5.11 shows the process arrangement of a Concat unit as it is used for instance to remove the sulfur components from the offgas leaving a non-

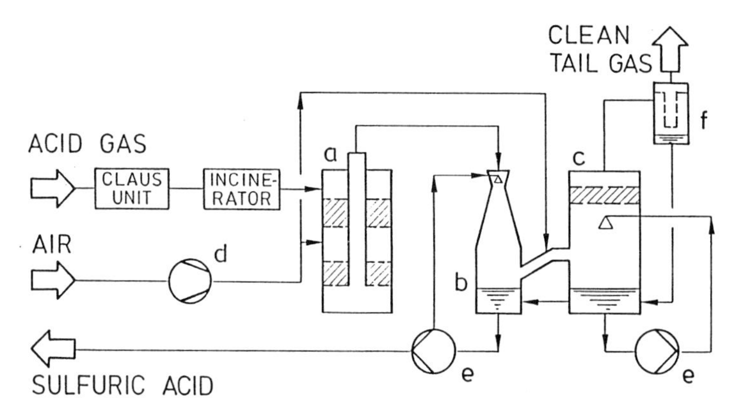

Fig. 5.11. Concat process. (*a*) Reaction tower; (*b*) scrubber; (*c*) condensing tower; (*d*) blower; (*e*) pump; (*f*) mist separator

selective coal gas desulfurization unit. This gas may contain some 5 % H_2S and COS together with 95 vol. % CO_2.

All sulfur components are substantially oxidized to SO_2 in a combustion chamber. The gas mixture containing the sulfur dioxide is routed to the Concat reactor containing the vanadium pentoxide catalyst. Here SO_2 as well as other residual sulfur compounds are oxidized to SO_3 by addition of air at temperatures between 400 and 450°C. Reaction temperature is controlled by quenching with air. 99.0 % of the sulfur feed leaves the reactor as sulfur trioxide. The humid gas leaving the reactor is treated with hot sulfuric acid in a hot condensing stage, and most of the SO_3 is obtained as sulfuric acid. The heat of reaction is removed by recycling and cooling of sulfuric acid.

From the Venturi washer, the gas is transferred to a condensing tower, where it is cooled by air to ambient temperature. The cooling air absorbs water vapour and effects a certain drying of the gas. Residual sulfuric acid vapours are condensed by spraying the gas with lean acid, which is recycled to the hot condensing stage.

A special filter is used to eliminate sulfuric acid mist from the gas/air mixture before the latter can be discharged to the atmosphere. As the gas mixture contains less than 200 vol. ppm of SO_2, it meets even the most stringent environmental requirements.

As in the Concat process, the wet catalysis principle is used also by a process developed jointly by *Topsoe* and *Snea*. This process differs from the Concat process by an additional acid concentration stage so that sulfuric acid concentrations up to 95 % can be reached. The SO_2 content of the offgas from this process, too, remains below 200 vol. ppm. Like the Concat process, the *Topsoe* process can also be used for treating tailgases from *Claus* units.

5.3.4 Tailgas Clean-up

The general name *tailgas clean-up* has been used for a number of processes developed over the past decade for an after-treatment of tailgases from *Claus* units. Even if a three-stage *Claus* system is used, the tailgas sulfur contents are still much higher than tolerated at most locations. In addition, some of the sulfur occurs in the form of H_2S for which maximum emission to the atmosphere is limited to 10 vol. ppm. If an acid gas containing 50 vol. % H_2S, 0.5 vol. % COS and 49.5 vol. % CO_2 is treated in a three-stage *Claus* unit, the tailgases from this unit still contain approximately 2 % of the raw gas sulfur content – 0.6 % in the form of SO_2 and 0.06 % in the form of H_2S. It is especially because of its high H_2S content, that such a gas needs to be cleaned up.

If the total sulfur content of such a tailgas can be tolerated and only the H_2S content has to be reduced to less than 10 ppm, *incineration* offers itself as the easiest way of tailgas clean-up. It is possible either in a *purely thermal* or in a *thermocatalytic system*. If thermal incineration is selected, the tailgases from the *Claus* unit are burnt together with fuel gas, i.e. the sulfur components of the tailgas are almost completely oxidized to SO_2 in an incinerator with a firebox temperature of 700–800°C. The CO formed in the *Claus* unit, too, is incinerated to residual contents of 100–200 ppm. Thermal incineration is used wherever the operating conditions cannot be assumed to be absolutely constant, i.e. in systems in which hydrocarbon or hydrogen sulfide breakthrough has to be anticipated whose incineration would lead to temperatures in the combustion zone which a catalyst could not survive. A thermal incinerator produces relatively little SO_3 and can easily handle also gaseous wastes containing fluorine or chlorine which would poison a catalyst.

Catalytic incineration at temperatures between 350 and 500°C should be considered wherever the cost of fuel gas is very high. It requires the tailgas flow rate and composition to be more or less constant. The catalysts used in this system consist of silica oxide doped with cobalt/molybdenum or aluminium oxide doped with copper/bismuth. The latter are characterized especially by the fact that they produce virtually no SO_3.The simplest way of cooling the hot incinerator offgases down to a temperature at which they can be discharged to the atmosphere through a conventional flue gas stack is mixing them with a sufficient volume of air. The air pressure may be raised to the slight over-pressure required for mixing by means of either the combustion air blower or of a separate blower. In most cases, however, especially in large plants, energy-saving considerations require the waste heat to be recovered from the flue gases. To this end, the flue gases are used either to preheat *Claus* tailgases and/or combustion air in order to save fuel gas, or to generate high- and medium-pressure steam. As the flue gases are cooled down, it is essential that the temperature at the heat exchange surfaces is not reduced below the sulfur dew point as this would either lead to severe corrosion or require the use of sophisticated materials. The pollutant contents in the incinerator offgases are more or less the same for both processes, except for the CO content which is clearly higher for the catalytic system.

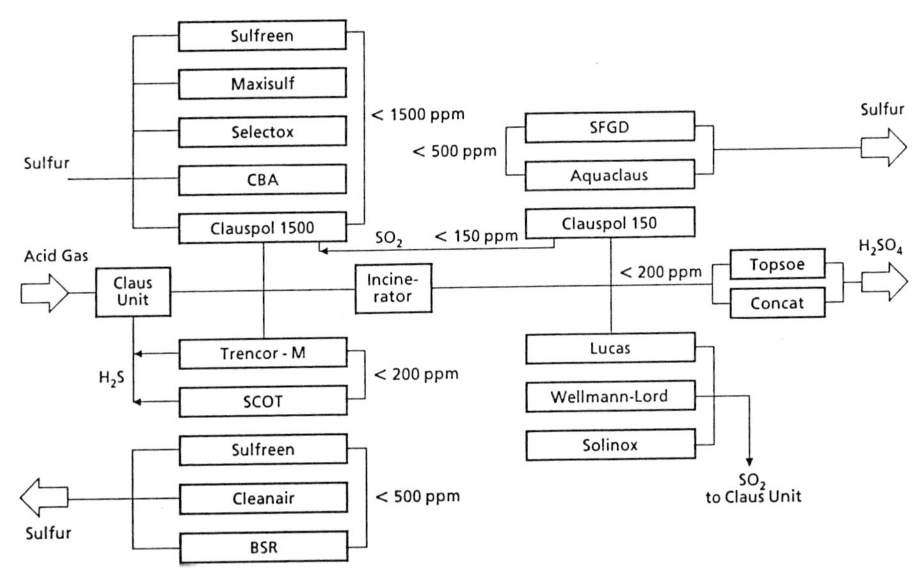

Fig. 5.12. Combination of *Claus* and tail gas clean-up processes

An overview over the numerous processes by which *Claus* tailgases can be genuinely treated to reduce sulfur emissions is given in Fig. 5.12. It shows the products resulting from the various processes and the achievable residual contents (H_2S + SO_2). As a detailed description of all the processes would go far beyond the scope of this chapter, only two processes which are typical for their respective groups will be dealt with here in greater detail while the others will only be mentioned briefly.

The LURGI *Sulfreen Process* is very similar to the *Claus* process and can be conveniently combined with it. Its major difference lies in the lower operating temperature which ensures that the sulfur recovered over the catalyst is simultaneously adsorbed by it.

Figure 5.13 illustrates that the *Claus* tailgases are fed to one of two cyclically operating reactors filled with highly active aluminium oxide catalyst. H_2S and SO_2 are reacted to elemental sulfur at temperatures around 130°C. The catalyst adsorbs this sulfur, and, when a certain maximum load has been reached, is thermally regenerated at approximately 300°C to desorb sulfur. During desorption, a stream of treated offgas is circulated through the loaded reactor by means of a blower. The loaded regeneration gas is then cooled and the sulfur recovered in a separator before the gas is recycled by the blower. The regenerated reactor is cooled down to the required operating temperature by means of a cold stream of treated offgas. The sulfur offgas leaving the Sulfreen reactor during its loading cycle is led to an incinerator where all sulfur compounds are converted to SO_2 except for a residual content of about 5 ppm H_2S. The Sulfreen Process can be used to achieve residual sulfur contents (SO_2 + H_2) of about 1 000 vol. ppm in the offgas.

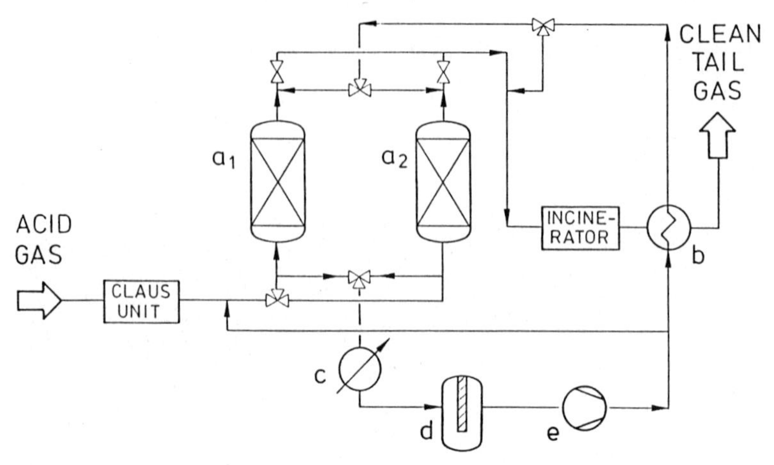

Fig. 5.13. Sulfreen process combined with *Claus* unit. (*a*) Reactor; (*b*) heat exchanger; (*c*) sulfur condenser; (*d*) separator; (*e*) blower

The *Scot Process* (SHELL *Claus* Off-gas Treating) developed by *SHELL Internationale Maatschappij* already in 1971 uses two steps:

- a hydrogenation stage to hydrogenate all sulfur components in the *Claus* tailgas,
- an absorption stage to absorb the resulting H_2S, desorb it from the wash liquor, and recycle it to the Claus unit.

The process arrangement is illustrated by Fig. 5.14. The tailgases leaving the Claus unit at temperatures around 140°C are heated to some 300°C by fuel gas combustion. As hydrogenation requires reducing gases, the fuel gas may

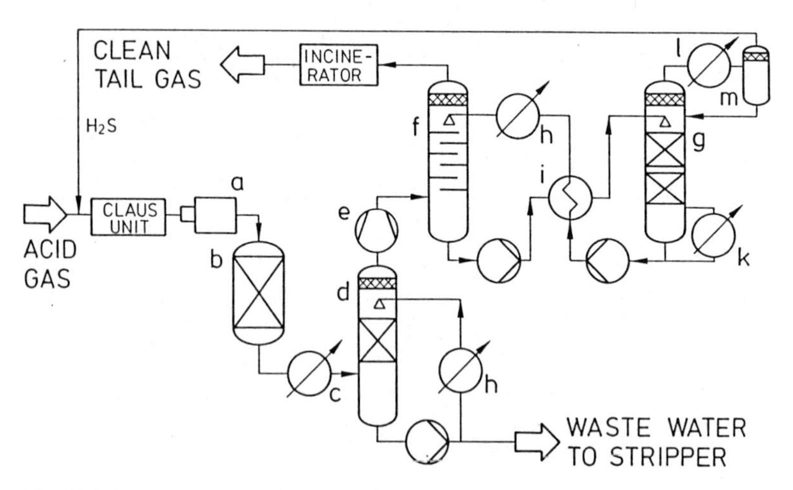

Fig. 5.14. Scot process combined with *Claus* unit. (*a*) Combustor; (*b*) reactor; (*c*) waste heat boiler; (*d*) quench cooler; (*e*) blower; (*f*) absorber; (*g*) regenerator; (*h*) cooler; (*i*) heat exchanger; (*k*) reboiler; (*l*) condenser; (*m*) separator

Claus sulphur recovery with SCOT tailgas clean up

consist either of gases containing hydrogen and carbon monoxide, or of low-boiling hydrocarbons which are burnt in a reducing atmosphere to produce H_2- and CO-rich gases. The gas mixture is then led through a reactor containing a cobalt-molybdenum catalyst to hydrogenate all sulfur components to H_2S. The two major reactions may be described by the equations

$$SO_2 + 3H_2 \rightleftarrows H_2S + 2H_2O$$

$$S_8 + 8CO \rightleftarrows 8COS.$$

The reaction continues almost to exhaustion, with residual SO_2 contents of less than 10 ppm.

The gas leaving the hydrogenation reactor is first cooled in a waste heat boiler to about 170°C and then in a quench cooler to approximately 35°C. The quenching operation uses circulating water from which the heat is removed in a cooler. Unlike the Sulfreen process described above, this process produces condensate as the gas is cooled below the steam dew point; this condensate has to be withdrawn from the water loop and fed to a stripper to remove H_2S. A blower delivers the almost dry gas from the quench cooler to an absorber to eliminate the hydrogen sulfide by means of an amine wash liquor.

As coabsorption of CO_2 leads to an increase in the size of the combined *Claus* and Scot system, it might be desirable to remove H_2S as selectively as possible. Selective amine solutions are therefore used, normally diisopropanolamine (Dipa), or methyl-diethanolamine(MDEA) if CO_2 contents are very high.

The treated offgas leaving the absorber contains between 200 and 500 ppm of residual H_2S, depending on how sophisticated the absorption and regeneration sections are designed. This residual H_2S is normally incinerated to SO_2 and discharged to the atmosphere. The spent wash liquor is heated up in a countercurrent to regenerated solution and then delivered to the top of the regenerator. The amine solution is regenerated with steam, while the expelled H_2S is cooled down in an overhead condenser and recycled to the *Claus* unit.

The Scot process is today considered to be one of the most flexible tailgas cleaning processes, primarily because the hydrogenation stage facilitates treatment of high COS and SO_2 contents and because fluctuations in the gas rate and concentrations which would otherwise force the *Claus* unit into a non-optimum operation are to a certain extent buffered off by the Scot unit.

All other processes, which are not dealt with in detail here, follow one of the three previously described principles:

- adsorption and desorption of elemental sulfur
- absorption of SO_2 or H_2S (the latter after previous hydrogenation) followed by desorption and recycling to the *Claus* unit or withdrawal of the product sulfur
- wet catalysis, producing sulfuric acid.

The first group includes – in addition to the Sulfreen process described in detail above – also the *CBA* (Cold Bed Adsorption Process) developed by *Amoco Pro-*

duction Co. It is similar to the Sulfreen Process but the adsorbers are regenerated not with treated offgas but with process gas from the outlet of the first *Claus* stage. The sulfur recovery rate is better than 98 %.

The absorption principle for sulfur components is used by the *Clauspohl 1 1 500 Process* of *Institut Français du Petrole* (IFP). Sulfur dioxide and hydrogen sulfide are absorbed from the *Claus* tailgases by *polyethylene glycol*. The sodium silicate catalyst in the solvent converts the sulfur components to elemental sulfur. The liquid sulfur is then recovered from the solvent. Continuous addition of sodium hydroxide and salicylic acid make up for catalyst losses. The offgases leaving the combined absorption/reaction zone have to be incinerated. The combination of a Clauspohl 1 500 unit with a *Claus* unit ensures a sulfur recovery rate of more than 98 %. Any COS and CS_2 leaving the *Claus* unit cannot be reacted in a Clauspohl unit and may therefore increase the SO_2 content of the clean offgas considerably.

IFP also developed the *Clauspohl 150 Process* by which the sulfur contents in the incinerator offgases can be reduced to less than 150 ppm. The process is based on the absorption of SO_2 in an aqueous ammonia solution followed by thermal decomposition of the sulfides to ammonia, water and sulfur dioxide, and reacting the latter to elemental sulfur in a Clauspohl 1 500 unit.

The *Trencor-M Process* developed by *Trentham Corporation* is similar to the Scot process, differing from the latter only in so far that a solution of monoethanolamine in organic substances (which are not disclosed) is used for absorption. SO_2 contents of less than 200 ppm in the incinerator offgases are reached by this process.

Like the Scot process, the *BSR* (*BEAVON* Sulfur Removal Process) developed by *R.M. Parsons Cie*. and *Union Oil Company of California* hydrogenates all sulfur components of the *Claus* tailgases to H_2S. The H_2S is then removed in a Stretford unit (see Sect. 2.3.3) to obtain elemental sulfur. The clean offgases from the Stretford unit contain less than 200 ppm residual sulfur and less than 5 ppm H_2S.

A variant of the BSR process is the *BSR/Selectox-I Process*. With this process, H_2S absorption in a Stretford unit is replaced by oxidation to elemental sulfur over an aluminium oxide catalyst termed *Selectox 32*. The offgases from the oxidation stage have to be incinerated and will then contain residual SO_2 contents around 1 000 ppm. Sulfur recovery with this process remains below 99 %.

Numerous other processes such as the *Stauffer-Aquaclaus Process*, the *Wellmann-Lord* and the *Cleanair Process* have been described in the relevant literature [5.5,8].

6. How to Supply Utilities to a Coal-to-Methanol Plant

It is not only for the sake of completeness but also because they account for a major portion of energy demand and capital investment costs that this description should include also those utility production and conditioning units which are commonly termed offsites. Most coal gasification plants need oxygen, and the units to produce this oxygen consume a great deal of energy and, because of their very special technology, are frequently included among the process units. A great deal of attention has therefore been attributed to air separation units used for the production of oxygen and nitrogen.

6.1 Air Separation Units

Oxygen can be produced by various routes. The two main components of air, i.e. oxygen and nitrogen, can be separated by *molecular sieves* adsorbing one of the two more rapidly than the other. *Permeation technology* is another method by which separation can be achieved.

For the large-scale plants covered here, only one technology is worth considering to produce the oxygen required for most coal gasification processes, namely *air separation* at cryogenic temperatures. The development of this process from the first successful liquefaction of air in 1895 to the technically mature and highly cost-effective large-scale plants of today which can provide as much as 60 000 m^3/h of pure oxygen is associated above all with the names of *Linde*, *Claude* and *Fraenkl*. Figure 6.1 shows the major parts of a high-pressure air separation unit which dates back already to the turn of the century and whose modern variants are still used today. It is based on the *Linde* principle of using the *Joule-Thomson* effect to cool down the air to its dew point in a countercurrent to the oxygen and nitrogen produced, in combination with the *Claude* principle of cooling gas by letting it work out in an expansion engine.

The air to be separated passes a filter and is then fed to a compressor equipped with intercoolers and separators to raise it to pressures between 50 and 200 bar. In a first heat exchanger, the air is cooled down to little more than the water freezing point and then fed to alternating adsorbers, one of which adsorbs water, carbon dioxide and such trace elements as acetylene and hydrocarbons while the other is regenerated with dry warm gas. The clean air is then led to the cryogenic section in which some of it is cooled by flashing it to approximately 6 bar in an expansion engine. Before it enters the bottom of the pressure column, it

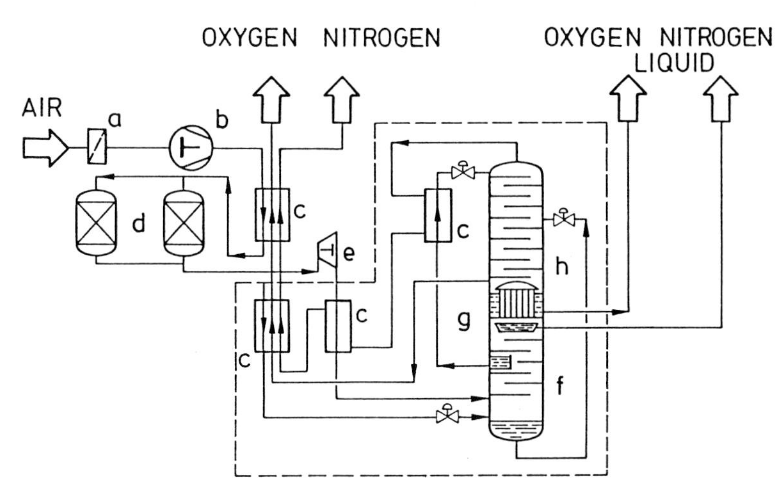

Fig. 6.1. High pressure air separation process. (*a*) Filter; (*b*) compressor; (*c*) (1–4) heat exchanger; (**d** adsorber; (**e** expansion mashine; (*f*) high pressure column; (*g*) main condenser; (*h*) low pressure column

is cooled down to little more than its dew point, partly cooled in a heat exchanger, and depressurized by way of a butterfly valve. Both streams are then fed to the pressure column of the separation unit.

Rectification takes place in a twin-column rectifier. The air fed to the pressure column is rectified to pure nitrogen which leaves the column over-head and is liquefied in the condenser. The major part of the liquid nitrogen is used as reflux to keep rectification going while a smaller percentage can be withdrawn as liquid product. The low-pressure column derives its reflux also from the pressure column. Liquid streams are withdrawn from the central and bottom part and flashed into the top column.

The column reflux is rectified in a countercurrent to rising steam to obtain pure oxygen which can be withdrawn from the condenser as a liquid or vapour. Nitrogen condensation in the pressure column is made possible by evaporating oxygen in the condenser. The impure nitrogen obtained at the head of the top column contains as much as 5 vol. % oxygen.

Flashing the air by a very high pressure differential leads to extensive chilling which ensures that a considerable part of the production can be withdrawn in liquid form. The energy demand of the high-pressure process is as much as 10 % higher than that of modern process modifications. Nevertheless, this systems is still used even today for smaller units in view of its very simple configuration. The broken line in Fig. 6.1 encloses those parts of the unit which are installed in the so-called “cold box” and insulated together.

The process bearing the name of *Linde-Fraenkl* is characterized by a low operating pressure and by using regenerators. The air is compressed to approximately 6 bar by means of a turbo-compressor and then cooled in a trickle cooler and fed to the cold box. First of all, it flows through one of the regenerators, which had been precooled by means of cold separation products, and is there cooled to close to its dew point, i.e. to approximately 100 °K. Next, it is fed to the separation unit to be separated in the previously described way. In this case, too, the chilling demand of the unit is provided by means of an expansion engine in the form of a turbine. In this respect it differs from the high-pressure process which uses reciprocating machines. The substances deposited on the packings (aluminium internals, rock packings) as the air flows through the regenerators, for instance water, carbon dioxide and other trace components of the air which are condensed at the prevailing temperatures, are discharged from the regenerators by the impure nitrogen. The pure oxygen and nitrogen, too, contribute to cooling the regenerator in the *regeneration cycle* although they flow through it in separate tube coils. The regenerators are switched over from one cycle to another by means of a switching machine.

Considering that not all the harmful or dangerous trace components of air, such as hydrocarbons, but above all acetylene, are condensed or frozen out in the regenerator, special adsorber loops to remove these substances are provided in the liquid oxygen system, where these substances would have a particularly negative effect.

The regenerators have in many modern plants been replaced by reversing exchangers in the form of plate heat exchangers whose heat exchange surfaces are designed for condensation and/or sublimation of H_2O, CO_2 and trace components. The same system is used also in low-pressure units with a superposed high-pressure loop in order to increase the output of liquid products. This becomes necessary whenever oxygen is to be used for gasification at high pressures of 60 bar or more and the plant user prefers liquid oxygen pumping to the use of highly sophisticated turbocompressors in spite of the fact that this is a disadvantage with respect to energy economy. The operation of the superposed high-pressure loops is based on the fact that some of the air or nitrogen is compressed to something like 50 bar and is then flashed again to provide the required additional refrigeration capacity.

Air separation according to the *medium-pressure route* has been gaining greatly in importance already over the past decade. As such plants include no switchable heat exchangers, neither of the product flows is required wholly or partly for flushing to discharge carbon dioxide and water. Hence, this process yields as much as 75 % of the ingoing air as pure products, whereas regenerative systems turn out no more than 55 % as a pure product. Air separation units based on this system can today be constructed for air capacities up to 310 000 m^3/h corresponding to 60 000 m^3/h of pure oxygen.

Figure 6.2 shows the process flow diagram for such a plant. Air is compressed to 7–8 bar, precooled in countercurrent heat exchangers and then cooled in a chiller to near the water freezing point before the condensate is drained. Thereafter, the air is further cooled by heat exchange to something like 180 °K, and the carbon dioxide, acetylene and other hydrocarbons are removed in other adsorbers. The air, which is now clean, is further cooled down in countercurrent heat exchangers and flashed in an expansion turbine. A slip stream of air

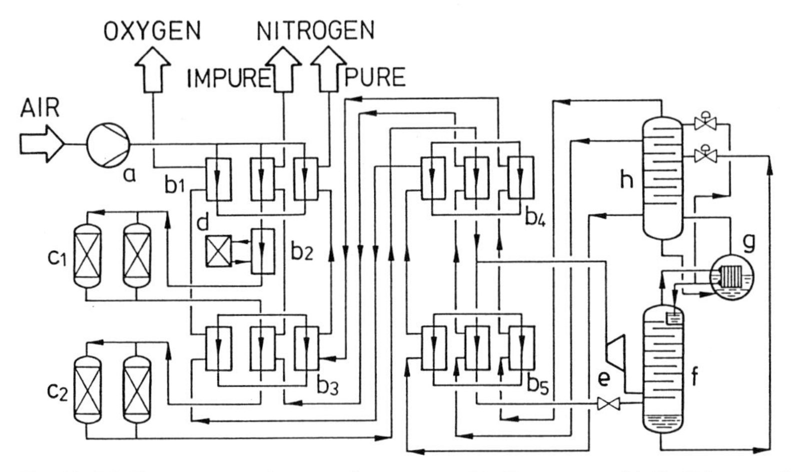

Fig. 6.2. Medium pressure air separation process. (*a*) Compressor; (*b*) (1–3) heat exchanger; (*c*) 1. dryer, 2. low pressure column

bypasses the expansion turbine to be flashed directly into the bottom of the pressure column. When the air has been separated as described above, all gaseous products are heated up to ambient temperature in a countercurrent to the air that needs to be cooled down.

Both the low-pressure and the medium-pressure processes can easily produce oxygen of more than 98 vol. % purity. The remaining 2 vol. % consist of about two thirds argon and one third nitrogen. If it is important that the synthesis gas should be virtually free from inert gases, an oxygen concentration of 99.8 vol. % can be reached with little additional expenditure. Some of the nitrogen can be exported with less than 50 ppm residual oxygen while the remaining nitrogen normally contains between 4 and 5 vol. % oxygen.

6.2 Steam and Electric Power

Whereas during the past century the chemical industry – unlike power stations – still used medium-pressure systems and only slightly superheated steam, the need for the best possible efficiency has meanwhile forced chemical plants, too, to use steam pressures as high as 120 bar and temperatures up to 500 °C. Such systems are cost-effective above all if the plants consume large amounts of energy, as in the case of coal-based methanol plants, whose total installed output – steam plus electricity – already reaches the order of small power stations.

Contrary to power stations which derive all the heat they need to generate and superheat steam from combustion, chemical plants often derive a considerable part of their entire thermal demand from process heat. Processes whose reactions are strongly exothermic, such as the methanol synthesis reaction, may in some cases produce steam of as much as 50 bar.

Some of this steam, for instance process steam for certain coal gasification processes or for CO shift conversion, is required at pressures up to about 50 bar, and there is normally a demand for low-pressure steam of 5–10 bar, and in most cases even for general heating steam of about 2 bar. The steam generated at high pressure (120 bar), which may partly come from the waste heat of high-temperature coal gasification plants, is superheated to 480–500 °C and then expanded in the turbines used to drive large machines. Process and heating steam is then bled from these turbines at a pressure and rate as required, and the remaining driving work is performed by condensation.

Frequently it will be appropriate to oversize the condensation stages of machines which have to be started up at a time when steam generated from process heat is not yet available, i.e. when the entire steam output has to be provided by fired steam boilers. This reduces the steam rate required to start up the plant and ensures that the steam boiler plant can be dimensioned so that it meets the normal continuous steam demand of the plant. However, it will be desirable as a rule to use bleeding and backpressure turbines which are not only much more cost-effective than condensation turbines, but which also improve the overall energy efficiency considerably owing to the combination of energy, process steam and heating steam.

The steam boiler plants are preferably fired with sulfur-free or low-sulfur fuels from the plant itself, for instance purge gas from methanol synthesis or flash gas from coal gasification and gas scrubbers. Wherever this heat supply is insufficient, it may be supplemented by firing natural gas, if available, or coal. The latter often requires complex desulfurization systems so that in certain cases even the production of low-sulfur fuel gas by coal gasification with air and subsequent desulfurization of the gas may be considered as an alternative to coal-fired systems.

A system particularly suited to provide both steam and power to a coal-to-methanol complex is the Circulating Fluid Bed Combustion Process together with conventional heat recovery and power generating systems. This technology, developed by LURGI in the seventies, combines high efficiency with low SO_2 and NO_x emissions. More than 90 % of the sulphur contained in the coal is captured in the combustion zone already, and due to low combustion temperature, less than 200 mg of nitrous oxides per m^3 of flue gas can be analysed. Coal burn-out is between 98 and 99 %, depending on the fuel.

Figure 6.3 shows the basic features of CFB. Coal and limestone are fed to the CFB-reactor. The coal is burnt with air at a temperature of approximately 850 °C. Primary air is fed through a grate located at the bottom of the reactor, secondary air is supplied part way up the combustor. Both streams of air not only provide the oxygen required for combustion but also keep the solids fluidized. The hot particles together with the gas produced leaving the rector are separated from each other in a recycling cyclone, the particles largely being recycled to the lower fluid of the reactor shaft. Most of the CFB plants are equipped with a fluid bed heat exchanger located at the level of the reactor bottom. In this heat exchanger, part of the hot particles are cooled and recycled to the reactor shaft.

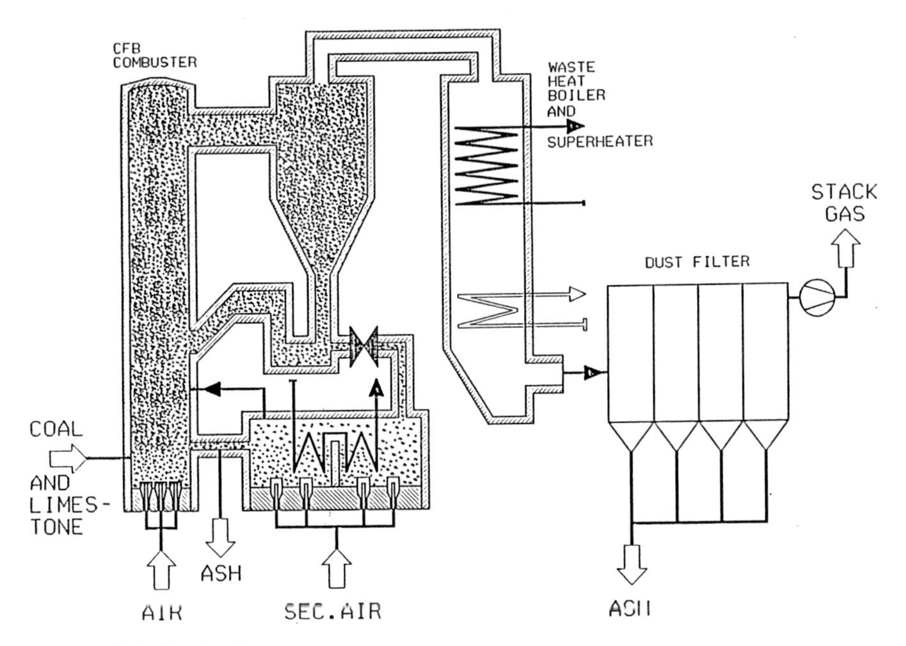

Fig. 6.3. CFB Combustion

The fluid bed heat exchanger not only helps to control combustion temperature but also serves to adjust the heat recovered for producing steam on the one side and to superheat it on the other side.

As Fig. 6.3 shows, the reactor shaft as well as the fluid bed heat exchanger are surrounded by *tubecages* in which high-pressure steam is generated. Superheating of this steam takes place in the convective section through which the gas together with some fly ash passes after leaving the recycling cyclone. Further heat recovery is achieved in the convective zone by preheating boiler feed water and combustion air. The flue gas finally passes either through electrostatic precipitators or through baghouses in which the fly ash is removed from the gas stream.

As the sulfur contained in the coal reacts with the limestone to form gypsum, the ash removed from the bottom of the reactor apart from the original coal ash contains only $CaSO_4$ and small amounts of lime and unburnt carbon. It can be harmlessly disposed of.

The power generation set combined with the CFB may be supplied with live steam of up to 150 bar and 535 °C. Process steam for process plants, primarily coal gasification, may be extracted from the turbine at convenient levels. Figure 7.4 shows an energy/heat diagram typical for a coal-based methanol plant.

Large machines are preferably driven by steam turbines because the latter are much more flexible in operation than an electromotor with constant speed and because the high startup currents of large motors often overtax the local power supply grids. Smaller machines, on the other hand, are invariably driven by electromotors. Considering the size of modern methanol plants, the term *smaller motors* is used here for all machines whose rating does not exceed 1–2 MW. This is advisable above all because such smaller machines are often installed in pairs (stand-by) or even in larger numbers to serve a single unit (air coolers) and because an electromotor is the driver that can most easily be started and stopped in such applications. Even if large machines are installed in pairs, for instance large process pumps, only the operating machine is as a rule driven by a steam turbine whereas the stand-by machine has an electromotor for quicker availability.

In some cases, plants have their own power supply system. This is important whenever frequent failures of external power supply have to be expected. As a coal-based methanol plant normally does not produce a surplus of usable heat, this electricity has to be provided via additional steam generation in the boiler plant. Wherever the public supply grid is sufficiently reliable, such in-plant power generation should normally be avoided because it does not really make the plant completely self-sufficient in energy. Either a stand-by unit would have to be provided for the in-plant power generator including the steam supply system or an emergency power supply from a public grid would have to be installed. Both alternatives are costly – the latter above all because not only the electricity imported during failures of the in-plant supply system has to be paid for but because the power companies charge a considerable commitment fee for keeping this power supply ready to cut in at any time.

A contribution to energy supply may come from the expansion of gases or liquids. Preferably, the turbines where this expansion takes place should be directly connected to an energy consumer. In large gas purification systems, considerable amounts of solvent quantities are brought up to the high pressure at which the absorber works and from there expanded into the regenerator which usually operates at nearly atmospheric pressure. If the high pressure pump is equipped with a turbine and the solvent flowing from the absorber to the regenerator is expanded through this so-called *reversed pump*, about 40 % of the total energy required for the high-pressure pump can be recovered. Likewise, it is possible to expand the purge gas from a methanol synthesis unit through a gas expander connected to the recycle gas compressor. After appropriate preheating – to avoid freezing of liquids at the expander outlet – some 70–80 % of the energy required for the recycle gas compressor can be recovered.

6.3 Water and Boiler Feed Water

Although modern plants are designed to export as little waste water as possible and attempts are made on the premises to recondition the waste water for cooling or boiler feed water service, almost all coal-to-methanol plants still have a considerable demand for raw water. Main consumers are cooling and boiler feed water supply systems, whereas other services such as potable water and water for general operating purposes account for only a small portion of the demand.

Few locations only can rely on clean well water. In most cases, the raw water has to be taken from rivers or even brackish waters. Almost all kinds of water contain one or more of the following impurities:

- dissolved or undissolved solids
- liquid pollutants
- gaseous pollutants.

The first category includes organic compounds such as top soil or similar material as well as inorganic compounds including all dissolved or suspended salts and minerals. The particle size of these impurities may range between 10^{-4} and 10^{-6} mm. Liquid pollutants mostly come from unnatural sources, consisting of mineral oil, gasoline or similar substances. Removing such substances from industrial water is a waste water treatment rather than water conditioning job.

Gaseous pollutants form part of all natural waters. They include not only inert or noble gases, nitrogen above all, which do not create problems in industrial applications, but also considerable quantities of dissolved oxygen and carbon dioxide which have to be removed in order to forestall serious damage to equipment and piping systems.

The first water conditioning stage is almost invariably a *mechanical cleaning stage*; it is absolutely necessary for all surface waters from rivers and lakes but is not needed for many well waters. Mechanical cleaning processes include:

- settlement processes by which suspended solids are made to settle on the bottom of the basin. Sometimes, this process is enhanced by chemical flocculants to shorten the retention time in the settling basins.
- filtration in order to obtain a virtually clear and colourless water which no longer contains any suspended solids. The filters used for this application are made of burnt diatomaceous earth or sinter metal powder, but also of plastic materials which are not always easy to clean. Very small particles have been retained successfully by precoated filters which gradually build up their filtering course during the filtration process itself. Such precoated filters are frequently replaced by gravel bed filters.

Mechanical cleaning is in many cases followed by a *chemical cleaning* stage, above all if the dissolved or undissolved solids cannot be eliminated by mechanical means alone. In such chemical stages, colloidal particles which normally repel one another owing to the identical charges of their surface ions are caused to agglomerate by adding ions of the opposite charge. These additives usually consist of activated salt solutions which form a flaky hydroxide deposit only when they get into contact with water. Particularly suitable for this application are aluminum sulfate or iron chloride. The flocculated deposit can then be removed by mechanical means. After this mechanical/chemical conditioning stage (if the latter is required), the conditioning process takes different routes for cooling water and boiler feed water. If the carbonate hardness of the mechanically treated raw water is too high, make-up water for the cooling loop can be obtained simply by softening it to an acceptable residual hardness. Make-up water for the boiler feed

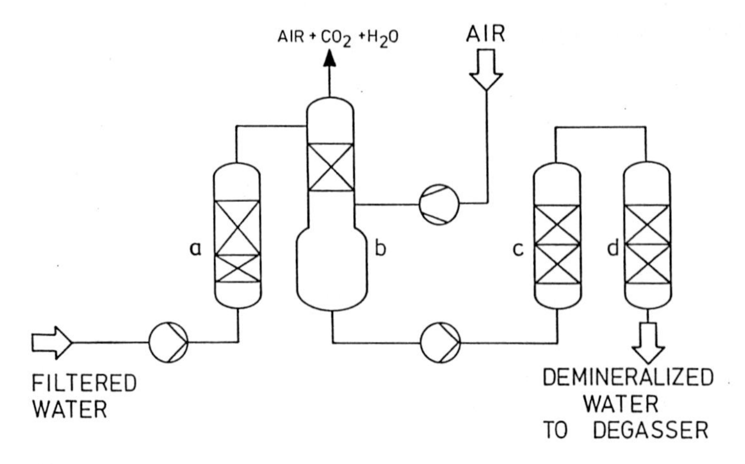

Fig. 6.4. Boiler feed water preparation. (*a*) Cation exchanger; (*b*) stripper; (*c*) anion exchanger; (*d*) mixed bed exchanger

water system, on the other hand, requires a number of additional conditioning steps, above all if high steam pressures impose very stringent re-quirements on the BFW quality.

Among all the earth alkaline metals contributing to the total hardness of the water, only calcium and magnesium are of practical importance for water conditioning. Calcium, which is most commonly found, is used also to define water hardness. One US hardness degree corresponds to 70 mg $CaCO_3$ per litre of water. Unlike the past, when water was decarbonized in a precipitation process by adding different chemicals, water is deionized today almost exclusively by means of ion exchangers. The process uses cation exchangers loaded with sodium ions which are exchanged for the calcium and magnesium ions in the water. Polystyrene resins are as a rule used in such cation exchangers. It may also be necessary in specific cases to reduce the iron and manganese contents of the raw water before it can be used as cooling water. To remove the iron, the water is saturated with air and the iron is retained by the filter in the form of iron hydroxide. A similar method is used to remove the manganese, but oxidation is normally supported by a manganese dioxide catalyst with which the gravel bed filter is coated.

How boiler feed water is prepared from mechanically cleaned water is illustrated by Fig. 6.4. The water is aerated and filtered to remove all solids and then pumped to the first vessel of the demineralization unit – the cation exchanger whose ion exchanger filling replaces the cations in the water by hydrogen. For calcium hydrogen carbonate, for instance, this may be described by the formula $H_2 + Ca(HCO_3)_2 \rightleftarrows Ca + 2H_2CO_3$. When the ion exchanger has been exhausted, it is regenerated by diluted hydrochloric acid which reacts with the metals on the ion exchanger resin to produce metal chlorides which can then be discharged while the ion exchanger itself is saturated with hydrogen. The hydrogen carbonate produced in the cation exchanger is decomposed with air in a downstream stripper and the resulting carbon dioxide is expelled.

Table 6.1. Boiler feed water specification*

Oxygen	(mg/l)	max.	0.02
Hardness	(mval/l)	max.	0.01
Total Iron	(mg/l)	max.	0.03
Copper	(mg/l)	max.	0.005
Total Carbonic Acid			
As CO_2	(mg/l)	max.	1.0
pH (at 20°C)		7 -	9.5
Silicic Acid	(mg/l)	max.	0.02
Permanganate Number	(mg $KMnO_4$/l)	max.	5.0
Oil	(mg/l)	max.	0.5

* As required for Steam Boiler operating at pressures from 80 bar up.

On leaving the stripper, the water to be deionized is fed to an anion exchanger to remove both the anions and any silicium that may be present in the water. One of the reactions which are typical for this process may be illustrated by the following formula: $H_2SO_4 + OH \rightleftarrows H_2O + SO_4$. The anion exchanger is regenerated with sodium hydroxide, discharging the SO_4 retained by the resin in the form of sodium sulfate while the ion exchanger is again saturated with OH ions.

The last vessel along the water desalination route, i.e. the mixed-bed filter which is occasionally also termed *the polisher*, serves as a common safety stage for the upstream cation and anion exchangers. it is used to remove the last cations and anions from the water. This mixed-bed filter is regenerated by applying the above two regeneration measures one after another.

After leaving the demineralization unit, the boiler feed water is heated to 95–100 °C and fed to the thermal degasifier on top of the boiler feed water tank. In this vessel, the BFW is stripped with steam to remove all traces of dissolved gases, above all of oxygen and carbon dioxide. Table 6.1 contains typical data to show the stringent BFW requirements when a boiler is operated at 80 bar. Clean steam condensate, for instance from turbine condensation units, normally requires no pretreatment before it can be fed to the thermal degasifier, whereas process effluents and steam condensates which may contain impurities have to run through some or all the stages of the demineralization unit. The remaining hardness of the boiler feed water is neutralized by adding phosphates to condition the BFW to the required pH; hydrazine (N_2H_4) is added to bind any "acids" which may still be present, above all oxygen, producing water and nitrogen as decomposition products.

6.4 Cooling Systems

The cooling systems required to remove the low- temperature process heat depend heavily on the plant location. The availability of water (wells, rivers, sea water) and the meteorological conditions are important factors for the selection of an optimum cooling system.

Air coolers are normally used to cool down process fluids to approximately 60 °C and for condensation units operating above such temperatures. These air coolers are at least as energy-efficient as water coolers over these temperature ranges and require some 2–3 kWh per GJ of discharged heat. Temperatures are controlled either by switching fans on or off, or by closing louvers to shade some of the cooling tubes. Air coolers require considerable space.

Water cooling systems may be of different types:

- open cooling loops
- closed cooling loops
- direct cooling by river or sea water.

The cooling systems which are most frequently used in chemical plants are *open-loop systems*. In such systems, the process heat is transferred to cold cooling water which in turn yields the absorbed heat to the ambient air in open cooling towers. The cooling effect in such systems results from evaporation of water. To enhance this effect, engineers have increasingly tended to use not only the natural draft resulting from the stack effect of these cooling towers, but to install large low-speed fans to increase the air throughput. This ensures that as much as 10 000 m^3/h of cooling water can be recooled in a single tower, corresponding to a heat dissipation to the atmosphere of 420 GJ per hour on the basis of the usual temperature differential of 10 °C between the hot and cold cooling water.

The cooled water is collected in a concrete basin underneath the cooling tower and pumped from there through the plant cooling systems. Since evaporation in the cooling tower consistently concentrates the salt in the cooling water, a small quantity of water – depending on the salt content of the make-up water and on the allowable salt concentration in the loop – has to be continuously withdrawn. This blowdown water and the water lost by evaporation have to be replaced by appropriately conditioned make-up water.

The cooling water has to meet the following requirements:

- non-corrosive
- free of algae
- free of solids.

Solids are removed at the water conditioning stage, if necessary by increased blow-down from the cooling water system. To prevent algae growth on the cooling surfaces which may detract from the cooling efficiency of the system, the cooling water is normally chlorinated. To do so, priority is normally given to batch chlorination at regular intervals as it is more reliable in killing algae growth and does not to the same extent lead to stress corrosion cracking in austenic steels as continuous chlorination. Carbon steel surfaces are normally passivated by suitable chemicals, mostly phosphates, to counteract corrosive effects from the cooling water and to prevent calcareous deposits.

Closed cooling loops are selected whenever

- the cooling water has to meet specific purity requirements
- make-up water is not available in the quantity or quality required for an open system
- meteorological conditions (very low temperatures) complicate the operation of open systems
- direct cooling with sea water is not advisable – although sea water may be available at appropriate temperatures – because of material problems (chlorine-induced corrosion of austenitic steels)) or fouling of heat exchanger surfaces (algae, sand).

Closed cooling loops are normally operated with boiler feed water or condensate. Air coolers or sea water may be used to discharge the heat from the loop. If sea water is used, the heat exchangers are normally made of titanium and twinned so that they can be cleaned without interrupting normal operation. As the sea water rate is very rarely limited, such exchangers can be operated with a very low temperature differential between the ingoing and the outgoing seawater – normally about 5 °C – in order to keep the expensive heat exchange surfaces to a minimum.

Larger water rates with a small δt are normally preferred also for *direct cooling* with river or sea water. Provided that the flow rate is kept within a relatively narrow range, carbon steel can still be used for sea water piping whereas the coolers themselves often require expensive alloys which are not always easy to handle, for instance admiralty metal or copper/nickel alloys.

A combination of two cooling systems may sometimes be found as well, for instance in plants which have both a number of smaller process coolers and large turbine waste steam condensers installed. Whereas the latter are operated with sea water wherever possible, the former are normally combined into a small open or closed loop.

6.5 Inert Gas

Nitrogen is normally used to inertize vessels and piping, for inert gas blanketing of methanol tanks, and for startup operations including catalyst reduction. Depending on the application, this nitrogen may contain more or less oxygen and carbon dioxide. The simplest way to obtain inert gas is to burn sulfur-free gases in conjunction with a CO_2 scrubber where necessary. If, however, the requirements regarding residual contents of oxygen or nitrous oxides are stringent, only air separation will do the job. Since all coal-based methanol plants use oxygen, the required inert gas is normally provided by the air separation unit. Residual oxygen contents of less than 50 ppm, meeting the most stringent requirements, can easily be obtained in this way. Liquid nitrogen storage facilities with powerful evaporators to produce large nitrogen quantities within a short time are provided to bridge startup and shutdown periods.

6.6 Instrument and Operating Air

Instrument and operating air can in most of the cases described here be obtained by way of the air compressor in the air separation unit. The instrument air is dried in switchable gel dryers at pressures between 6 and 8 bar to a water dew point (referred to atmospheric pressure) which is clearly lower than for the lowest local temperatures, in order to prevent steam condensation in piping, instruments and control valves. Sufficient air to shut down the plant in case of malfunction is stored in an air receiver of appropriate capacity. The air compressor of the air separation unit is almost in all cases supplemented by a stand-by instrument air compressor which normally produces only the air rate required to start up the plant.

Operating air can simply be taken from the air compressor in the air separation unit. It need not be dried nor stored.

6.7 Other Offsites

A grass roots plant needs a whole lot of other offsites in addition to the utility production or conditioning units described in this chapter; some of these offsites form part of the infrastructure, others are associated with administration, safety and maintenance. The list below can no doubt be enlarged in specific cases:

- coal supply
- tankfarm
- loading and unloading facilities
- factory railway system
- road system
- electrical substation
- fire-fighting facilities
- sewage system
- communication systems
- buildings and equipment for administration, amenities, laboratories and workshops.

Construction and operation of these facilities normally make a considerable contribution to the total investment and in some cases have an influence on energy demand as well. They, too, need therefore to be appropriately considered at the design stage.

7. What Could a Methanol Plant Look Like?

The way of reasoning required to arrive at an optimum plant concept under the given circumstances may best be impressed upon the reader by presenting and discussing a complete methanol plant producing 5 000 tpd of methanol from Dakota lignite.

For this purpose, the following aspects are of prime importance:

- technical suitability and optimum profitability of the gasification process for the type of coal to be used,
- configuration of the gas treatment and conditioning processes and of the offgas and waste water treatment systems with a view not only to optimum technology and economics but also to strict conformance with environmental requirements,
- integration of the individual units into a common energy and heat concept to an extent that is economically justifiable without neglecting the possibility of operating the units largely independently,
- selection of industrially proven process technology.

If the LURGI Low-Pressure Methanol Process discussed in this paper is replaced by another methanol process, the pros and cons of the individual process steps may of course be different.

7.1 Methanol from Lignite – Application of the LURGI Pressure Gasification Process

Going through the above evaluation criteria in a systematic manner, the first question will be: Why should medium-temperature pressure gasification be used for the production of methanol syngas from lignite?

Here are a few arguments that speak in favour of doing so:

- The LURGI dry-bottom gasifier can handle virtually all types of coal irrespective of their ash contents and nearly irrespective of their water contents. Processes which require the coal to be ground down and slurried before it can be fed to the gasifier are at a disadvantage economically because lignites typically have a high natural water content and readily absorb additional water.

- Oxygen consumption of the dry-bottom gasifier is particularly low, because the countercurrent operation does not require the steam and volatile coal components to be heated to the reaction end temperature, and because the lignites are highly reactive.

The raw gas from coal can best be cleaned up by the Rectisol Process for the following reasons:

- Methanol as a wash liquor can remove all undesirable gas components, starting from hydrocarbons that can be condensed under normal conditions through carbon dioxide and all sorts of sulfur components down to impurities that occur only in traces and sometimes even in the ppm range.
- As the gas is obtained at ambient temperature and a pressure of 30–40 bar and has a relatively high CO_2 content, physical wash processes are more suitable than others.
- The max. residual sulfur content of < 0.1 ppm required for methanol synthesis is reached by a single cleaning stage.
- The sulfur content of the acid gas can be sufficiently increased by an appropriate scrubber design so that the gas flow to the Concat unit, in which the H_2S and COS from the raw gas are recovered in the form of sulfuric acid, is kept to a minimum. The sulfur content in the CO_2 offgas can be adjusted to conform to environmental requirements.

The liquid products such as tar, oil, naphtha and phenols, which are obtained as the raw gas from coal is cooled, are converted by partial oxidation (Sect. 2.4) into high-CO and high-H_2 gas. This route was chosen for the following reasons:

- Although these liquid products, if treated and sold in a suitable place, might yield a higher revenue than methanol, it may be questionable whether marketing these relatively small quantities of liquid hydrocarbons at the location of large methanol-from-coal plants would make economic sense in view of transportation and storage cost.
- The plants to treat, store and ship the liquid products would require considerable additional capital investment and make plant operations more complex.
- The only such liquids that are exported from the plant are ammonia and sulfuric acid. The former, because it is pure already when it is recovered from the gas liquor (Sect. 5.1) and need not be further treated, and the latter because the sulfur content of the coal cannot be used otherwise in the plant. The exported ammonia represents a loss of almost 1 % in terms of hydrogen converted in the methanol plant or of methanol yields.

The waste water from the pressure gasification of coal was precleaned by a combination of the Phenosolvan and the CLL process, a combination which seemed reasonable because

- the Phenosolvan Process cannot only preclean this waste water enough for a normal biological stage to handle it without problems, but also enables the valuable substances to be recovered from the water in the form of raw phenol which can either be used as boiler fuel – with appropriate credit for its calorific value – or processed to marketable products.

- the CLL unit in conjunction with a Phenosolvan unit makes it possible to obtain marketable ammonia on the one hand, and a virtually ammonia-free mixture of H_2S and CO_2 on the other which can then be conveniently processed in a Concat unit.
- no experience with other processes is available for this application.

Whereas Sect. 2.4 describes how the purge gas from a plant producing methanol alone is treated by autothermal reforming, an investigation was made alternatively also into co-production of methanol and SNG (Substitute Natural Gas). Provided that either suitable consumers or a natural gas pipeline exist, this co-production appears reasonable because it leads to a considerable increase in the overall efficiency of the plant, i.e. the ratio of heat output to heat input. In the present case, this efficiency rises from 52.7 % for methanol production to 59.8 % if methanol and SNG are co-produced.

7.2 Description of the Process Concept

Figure 7.1 shows the block diagram of a methanol-from-lignite plant.

Run-of-mine lignite is precrushed and screened as required in a coal preparation unit. The particle size of the feed to the dry-bottom gasifiers should not exceed 100 mm, nor should it fall short of 4 mm to maintain optimum gasifier efficiency. The feed consists of North Dakota lignite with properties as shown in Table 7.1.

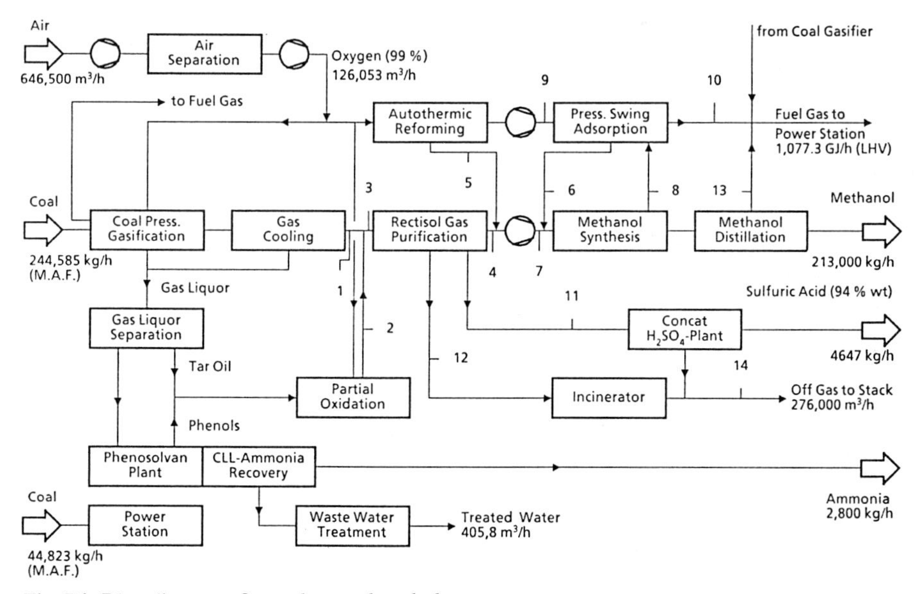

Fig. 7.1. Bloc diagram of a coal to methanol plant

Table 7.1. Properties of North Dakota lignite

Prox. Analysis:		
Moisture	30.20	% wt.
Ash	7.43	% wt.
Volatile Matter	30.55	% wt.
Fix Carbon	31.82	% wt.
HHV:	18,783	kJ/kg
Dry, ash-free Basis:		
C	74.89	% wt.
H	5.27	% wt.
N_2	1.14	% wt.
Cl	0.04	% wt.
S	0.60	% wt.
O_2	18.06	% wt.
HHV:	30,081	kJ/kg

The gasification unit consists of 6 LURGI dry-bottom gasifiers – no stand-by gasifier is required because of the large number of operating ones – and can gasify some 392 tons of lignite per hour, corresponding to 244 585 kg/h (m.a.f.), with oxygen and steam at a pressure of about 30 bar. The necessary oxygen – 87 653 m^3/h of 99 % purity – comes from a medium-pressure air separation unit and is compressed to 35 bar in a turbocompressor. 355 400 kg of steam of 38 bar and 450°C are taken from a grid which draws its supply partly from a power station and partly from the process units. An additional 46 700 kg/h of steam produced in the shell cooling system are fed to the gasifier.

This system produces 495 112 m^3 of raw gas (dry) per hour with the composition as shown in the material balance in Table 7.2. The gas is cooled to ambient temperature in several stages, producing some 228 tons of 7.5 bar steam per hour in a low-pressure waste heat boiler. The resulting condensates – hydrocarboneous and aqueous – are taken to the tar and oil separator as described in Sect. 1.3 and separated according to their density into tar, oil, and naphtha fractions and gas liquor. The dust-laden highly viscous tar is recycled to the gasifiers, while the gas liquor is pretreated in the CPI separator and gravel bed filter unit as illustrated by Fig. 5.7 and then fed to the phenol unit to be dephenolated by means of di-isopropyl ether as an extraction agent. The raw phenol together with the liquid hydrocarbons from the tar and oil separator and the gas naphtha from the Rectisol prewash stage, through which the raw gas flows first, is passed on to the single-train partial oxidation unit to be reacted at about 1 500°C and a pressure of 30 bar.

The reformed gas from this unit – a glance at the material balance shows that it contains as much as 13.0 % of the CO and H_2 found in the product syngas later – is also cooled down to ambient temperature, using most of its sensible heat

Table 7.2. Dry gas balance methanol

Gas. Comp. [% mol]	1 Crude Gas ex Coal Gasific.	2 Crude Gas ex POX	3 Gas to Rectisol	4 Gas ex Rectisol	5 Gas ex Autoth.Ref.	6 Hydrogen ex PSA	7 Synthesis Gas
CO_2	31.56	4.63	28.23	1.86	9.79		3.83
CO	15.42	46.42	19.25	26.74	23.51		24.45
H_2	40.59	48.24	41.53	57.45	62.32	100.00	61.02
CH_4	11.09	0.30	9.76	13.34	2.18		9.70
C_nH_m	0.92		0.81	0.28			0.19
H_2S + COS	0.20	0.04	0.18	<0.1 ppm			
SO_x							
N_2 + A	0.22	0.37	0.24	0.33	2.20		0.81
O_2							
CH_3OH							
[m^3/h]	495,111	69,812	564,923	406,058	154,911	31,010	591,893

Gas. Comp. [% mol]	8 Purge Gas to PSA	9 PSA Off-Gas to Autoth.Ref.	10 PSA Off-Gas to Fuel	11 Gas to Concat	12 Off Gas ex Rectisol	13 Fuel Gas ex Distil.	14 Off-Gas to Atmosph.
CO_2	6.06	8.03		91.45	96.20	28.45	60.4
CO	6.15	8.17		0.14	0.12	2.70	
H_2	39.87	20.11		0.96	0.88	8.11	
CH_4	43.54	57.86	As	1.18	0.53	38.81	
C_nH_m	0.89	1.18	in	1.31	2.27	0.80	
H_2S + COS			9	4.92			< 2 ppm
SO_x							<50 ppm
N_2 + A	3.49	4.63		0.04		1.13	38.7
O_2						20.00	0.9
CH_3OH							
[m^3/h]	125,359	72,768	21,581	20,768	138,183	8,273	276,000

to generate high-pressure steam (120 bar). This cooled and virtually soot-free gas is added to the gasification raw gas downstream of the Rectisol prewash unit and the combined gas flows are then taken to the Rectisol gas treatment unit.

Since the low ash contents of the feed result in extremely little fouling of the gasifiers – less than 0.5 % is quite normal –, a soot recovery and recycling system cannot be justified on economic grounds. Instead, the soot-laden water is discharged to the waste water system of the complex.

It may be worth noting here that partial oxidation of liquid hydrocarbons might also be replaced by a slurry gasification process (Sect. 1.3). This would enable the percentage of coal fines of less than 4 mm particle size – which in this typical case is restricted to be used in the power plant and accounts for slightly less than 20 % – to be increased almost arbitrarily and to add both the liquid hydrocarbons and the raw phenol to the coal slurry and gasify them together. However, since such a unit would introduce a carbon monoxide surplus in the system – the stoichiometric ratio of the methanol syngas would decrease to less than 2.0 – a CO shift conversion unit would have to be installed in one of the gas streams. The gas leaving the autothermal reformer would be the most suitable as it is dust-free, soot-free and sulfur-free and already contains the steam required for the shift reaction.

The absorber of the single-train Rectisol unit, in which the main wash stage is followed by a trimming stage, reduces the sulfur content of the gas – essentially H_2S, but also COS, mercaptanes and thiophenes – to 0.1 vol. ppm (H_2S + COS) and the CO_2 content to approx. 1.9 vol. %. The configuration of the Rectisol unit is similar to that shown in Fig. 2.9. Some 1 150 m^3/h of wash liquor are circulated to the absorber, of which about 660 m^3/h are pumped to the main wash stage after most of the CO_2 has been removed in three subsequent flashing stages.

The flash gas from the first stage, containing most of the co-absorbed hydrogen, carbon monoxide and methane, is recycled to the absorber to minimize the losses of valuable gases. In the second stage, the gases are flashed into the reabsorber bottom in which all CO_2 that is to be discharged to the atmosphere is washed once more with heat-regenerated sulfur-free methanol to reduce its H_2S content to something like 10 vol. ppm. The last flashing stage operates under slight vacuum. The virtually CO_2-free but still sulfur-laden methanol from the last flashing stage is for the most part pumped to the absorber with a smaller quantity going to the hot regenerator in which the methanol is heat-regenerated by indirect steam-heating to expel virtually all the sulfur.

The reabsorber ensures that all the sulfur entering the Rectisol unit is concentrated in about 10 % of the CO_2 absorbed from the raw gas. The acid gas has an H_2S content of 5 vol. %. The heat-regenerated methanol is fed to the absorber trimming stage.

The purified gas from the Rectisol unit is mixed with the gas from the purge gas reforming unit and with the hydrogen separated from the purge gas in the PSA unit. The mixture so obtained – some 592 000 m^3/h – constitutes the finished syngas with a stoichiometric ratio of about 2.02. A turbocompressor driven by a steam turbine compresses it from 24.5 to 73 bar before it is delivered to the synthesis loop where the carbon oxides and hydrogen are reacted to methanol at a pressure of about 70 bar.

The synthesis loop is a single-train unit. However, as the considerable size of the equipment might lead to difficulties with purchasing, transportation and erection, the system was split up into two parallel boiling-water-cooled tubular reactors and two gas/gas heat exchangers. Most of the heat of reaction is used to generate 40 bar steam which is collected in a single steam drum connected with both reactors and then delivered into the grid. Although the loop contains an extremely high proportion of inert gases – methane, nitrogen and argon account for almost 50 % of the recycle gas –, as much as 95 % of the CO and 66 % of the CO_2 in the feed are reacted. The raw methanol from the synthesis loop is taken to the methanol distillation section to be processed to grade AA methanol in a three-column system as shown in Fig. 4.2. The dissolved gases and low-boilers that are expelled overhead in the prerun column are fed to the fuel gas grid. Methanol distillation, too, takes the form of a single-train unit.

The purge gas which is withdrawn from the synthesis loop to limit its inert gas content is fed to a pressure swing adsorption unit where some 80 % of the hydrogen are removed and added to the purified gas downstream of the Rectisol unit as described above. 20 % of the remaining gas, which contains about 44 % methane, are discharged to the fuel gas grid. This gas volume contains virtually all the nitrogen and the entire argon that had entered the syngas together with the coal and the oxygen needed for gasification. The true inert gas content – methane is considered an inert gas only for the purposes of the synthesis loop – can thus be kept down to a tolerable level. The remaining 80 % of the residual gas from the PSA unit are preheated to about 450°C and then catalytically reformed with steam and oxygen in an autothermal reactor (Sect. 2.4) to obtain a high-CO and high-H_2 gas which contains absolutely no sulfur and therefore requires no further treatment but can be cooled and added directly to the purified gas from the Rectisol unit. As in the Partial Oxidation Unit, most of the sensible heat of the reformed gas leaving the autothermal reformer is used to generate high-pressure steam.

This leaves us with the question what happens to the dephenolated gas liquor, the acid gas and the CO_2 offgas from the Rectisol unit, and where the plant complex gets its electricity, steam and other utilities from.

The dephenolated gas liquor is first treated in a combined Phenosolvan/CLL unit (Sect. 5.1) to expel the acid gas and then recover 99.99 % pure liquid ammonia as illustrated in Fig. 5.5. The water leaving the total stripper goes to a biological treatment stage and is then aftertreated by means of activated carbon to serve as make-up water for the cooling loop and possibly also for the BFW loop.

The acid gas from the Rectisol unit and the CO_2 and H_2S removed in the CLL unit are jointly fed to a Concat unit to convert the sulfur components of the gas into 94 wt. % sulfuric acid (Sect. 5.2). With very few exceptions, the Concat offgas containing less than 1 ppm of H_2S and less than 100 ppm of SO_2 can be discharged to the atmosphere everywhere in the world without any af-

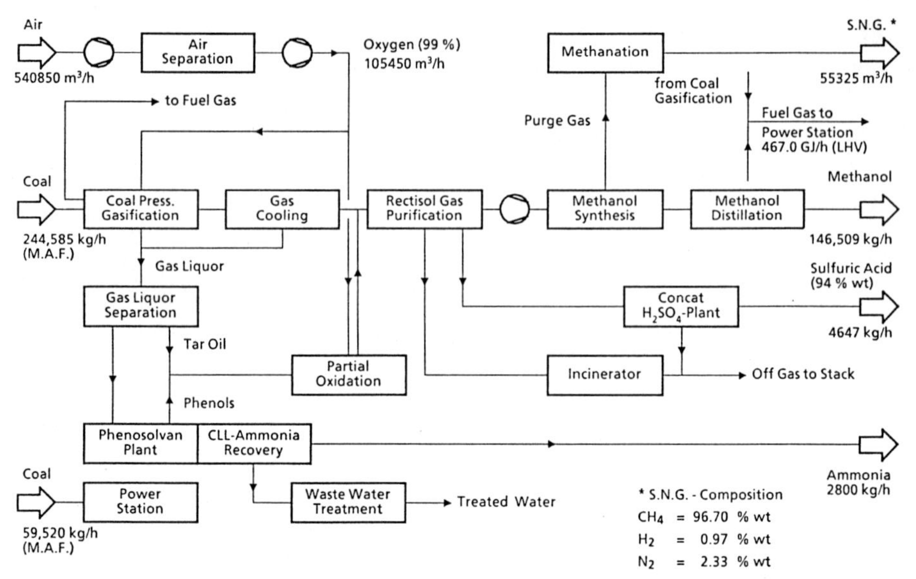

Fig. 7.2. Bloc diagram of a coal to methanol plus SNG plant

tertreatment or incineration. The CO_2 offgas from the Rectisol unit, however, has to be incinerated – not because of its H_2S content of less than 10 ppm which normally satisfies environmental provisions, but because its CO content of more than 0.1 vol. % and its hydrocarbon content of more than 2 vol. % are many times higher than what most industrialized countries regard as acceptable. A calculation on the basis of the material balance proves that only 18 kg of the almost 1 470 kg of sulfur entering the process plant together with the coal are emitted, i.e. the sulfur recovery rate without considering the power plant reaches almost 98.8 %.

The diagram of a plant that produces substitute natural gas (SNG) in addition to methanol is shown in Fig. 7.2. The entire purge gas stream from methanol synthesis is delivered to a methanation unit where the hydrogen and carbon oxides are reacted at a nickel catalyst to yield methane and water according to the following formulae

$$CO + 3H_2 = CH_4 + H_2O.$$

$$CO_2 + 4H_2 = CH_4 + 2H_2O$$

In order to ensure the necessary minimum hydrogen content in the SNG, the coal gasification unit is operated at a slightly higher H_2O/C ratio, as shown in Table 7.3. The purge gas rate leaving the methanol synthesis unit – some 125 000 m^3/h if methanol alone is produced – decreases to 80 000 m^3/h, and volume contraction during methanation finally produces 55 325 m^3/h of SNG with a net calorific value of 38 750 kJ/m^3. The methanol output decreases to 146 509 kg/h. The reduced export of fuel gas to the power station changes the utility balance signif-

Table 7.3. Dry gas balance Methanol + SNG

Gas Comp. [% mol]	1 Crude Gas ex Coal Gasific.	2 Crude Gas ex POX	3 Gas to Rectisol
CO_2	32.10	4.63	28.23
CO	14.49	46.42	18.41
H_2	41.06	48.24	41.94
CH_4	11.01	0.30	9.70
C_nH_m	0.92		0.81
H_2S + COS	0.20	0.04	0.18
N_2 + A	0.22	0.37	0.24
[m^3/h]	496,596	69,812	566,408

	4 Gas ex Rectisol	5 Purge Gas to Methanation	6 SNG
CO_2	1.85	2.89	] <10 ppm
CO	25.72	5.35	
H_2	58.47	29.46	0.97
CH_4	13.34	59.58	96.70
C_nh_m	0.29	1.26	
H2S + COS			
N_2 + A	0.33	1.46	2.33
[m^3/h]	406,720	87,942	55,325

icantly (in the PSA offgas section, less fuel gas from methanol distillation), so that some 33 % more coal has to be burnt in the power station. Nevertheless, the overall efficiency increases from 52.7 to 59.8 %.

An overview over the utilities required in the plant as a whole is given in Fig. 7.3. It can be seen that the only inputs to the plant are coal and raw water, i.e. the plant is self-sufficient regarding utilities, except for the chemicals required above all for BFW conditioning, and the lubricants and catalysts.

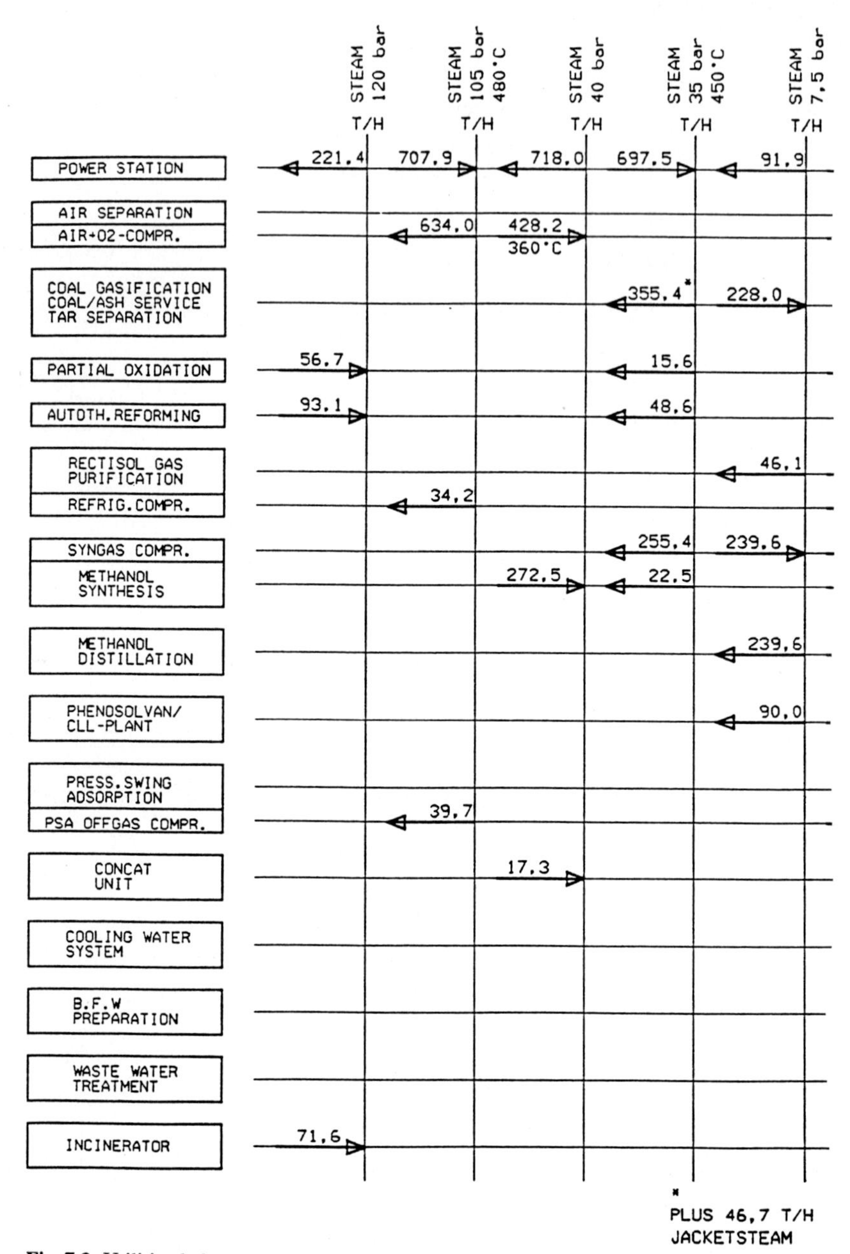

Fig. 7.3. Utilities balance

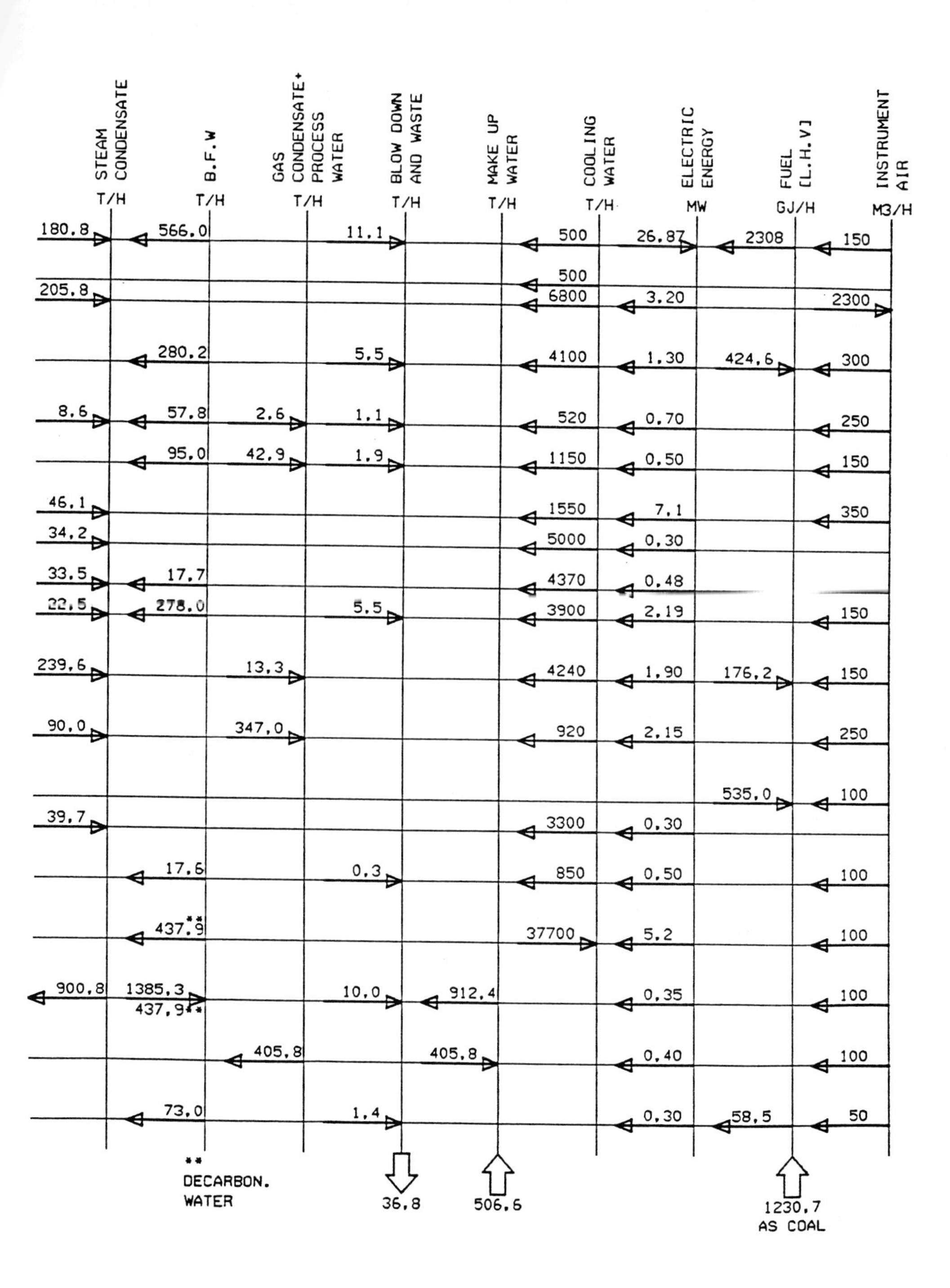

STEAM CONDENSATE T/H
B.F.W T/H
GAS CONDENSATE+ PROCESS WATER T/H
BLOW DOWN AND WASTE T/H
MAKE UP WATER T/H
COOLING WATER T/H
ELECTRIC ENERGY MW
FUEL [L.H.V] GJ/H
INSTRUMENT AIR M3/H
180,8 566,0 11,1 500 26,87 2308 150
500
205,8 6800 3,20 2300
280,2 5,5 4100 1,30 424,6 300
8,6 57,8 2,6 1,1 520 0,70 250
95,0 42,9 1,9 1150 0,50 150
46,1 1550 7,1 350
34,2 5000 0,30
33,5 17,7 4370 0,48
22,5 278,0 5,5 3900 2,19 150
239,6 13,3 4240 1,90 176,2 150
90,0 347,0 920 2,15 250
535,0 100
39,7 3300 0,30
17,6 0,3 850 0,50 100
437,9** 37700 5,2 100
900,8 1385,3 437,9** 10,0 912,4 0,35 100
405,8 405,8 0,40 100
73,0 1,4 0,30 58,5 50
** DECARBON. WATER
36,8
506,6
1230,7 AS COAL

That part of the steam generated in the process units which is not directly used for process or heating purposes goes to the power station where it is superheated together with the additional steam generated there for energy supply, process and heating. The power station also provides the electricity needed in the plant, and an integrated BFW deaeration unit supplies fully demineralized and deaerated water.

Figure 7.4 shows by a simplified diagram how the process units are supplied with process and heating steam and how the three steam levels – 105, 35 and 7.5 bar – are associated with the major energy consumers.

The heat required to generate and superheat steam is provided by a furnace plant using the CFB principle (Chap. 6). As shown in Fig. 7.3, this plant derives 1 230.7 GJ/h from coal and 1 077.3 GJ/h as fuel gas from the process units. Predominantly coal is burnt in the steam generator section, which consumes some 80 % of the total heat input, while the superheater section relies exclusively on sulfur-free fuel gas. An appropriate design of the steam generator and superheater sections ensures that the sulfur-free fuel gas is fed to that part of the equipment in which coal combustion is largely completed and the flue gas has already been desulfurized with limestone. Assuming that the coal gas is desulfurized at a rate of 95 % and that the flue gas originates from coal and fuel gas roughly in proportion to their respective heat inputs, the flue gas to the stack will contain approx. 40 mg of SO_2 per m^3 and its NO_x content will be about 100 mg/m^3. In the present example, about 18 kg of sulfur are emitted from the power station. The overall sulfur recovery rate thus reaches 98.0 %. This is an extraordinarily high figure in view of the low sulfur content of the feed coal.

Since the sulfur content of the process offgases is completely unrelated to the sulfur content of the coal, and that of the flue gases from the power station is unrelated to it over a wide range, the plant concept described here would lead to a sulfur recovery of 99.6 % if a feed coal with e.g. 3 wt. % sulfur were used.

Figure 7.4 illustrates also how power generation is integrated into the overall energy concept. Slightly less than 27 MW are generated by expanding superheated HP steam in a back-pressure condensation turbine. Plant power generation is normally synchronized with a public grid to ensure that power generation and consumption can be balanced to a certain extent without changing the duty of the plant generator. This is important above all when the plant is started up or shut down or during transitional states.

The BFW preparation unit is associated with the power station. It is a full demineralization unit producing a BFW quality as required for a boiler operating at 120 bar. The power station service pump is driven by a condensation turbine running on 120 bar steam. The stand-by pump is driven by an electric motor, which can be automatically started up more easily than a steam turbine.

The BFW preparation unit is designed in such a way that it can supply also deionized water for cooling water make-up. Figure 7.3 shows that only about 55 % of the water required for BFW preparation or to make up for cooling water losses have to be imported. Cooling water is essentially needed for the large machines with their associated steam condensers and for the trim coolers in the process units. The large process coolers are designed as air coolers.

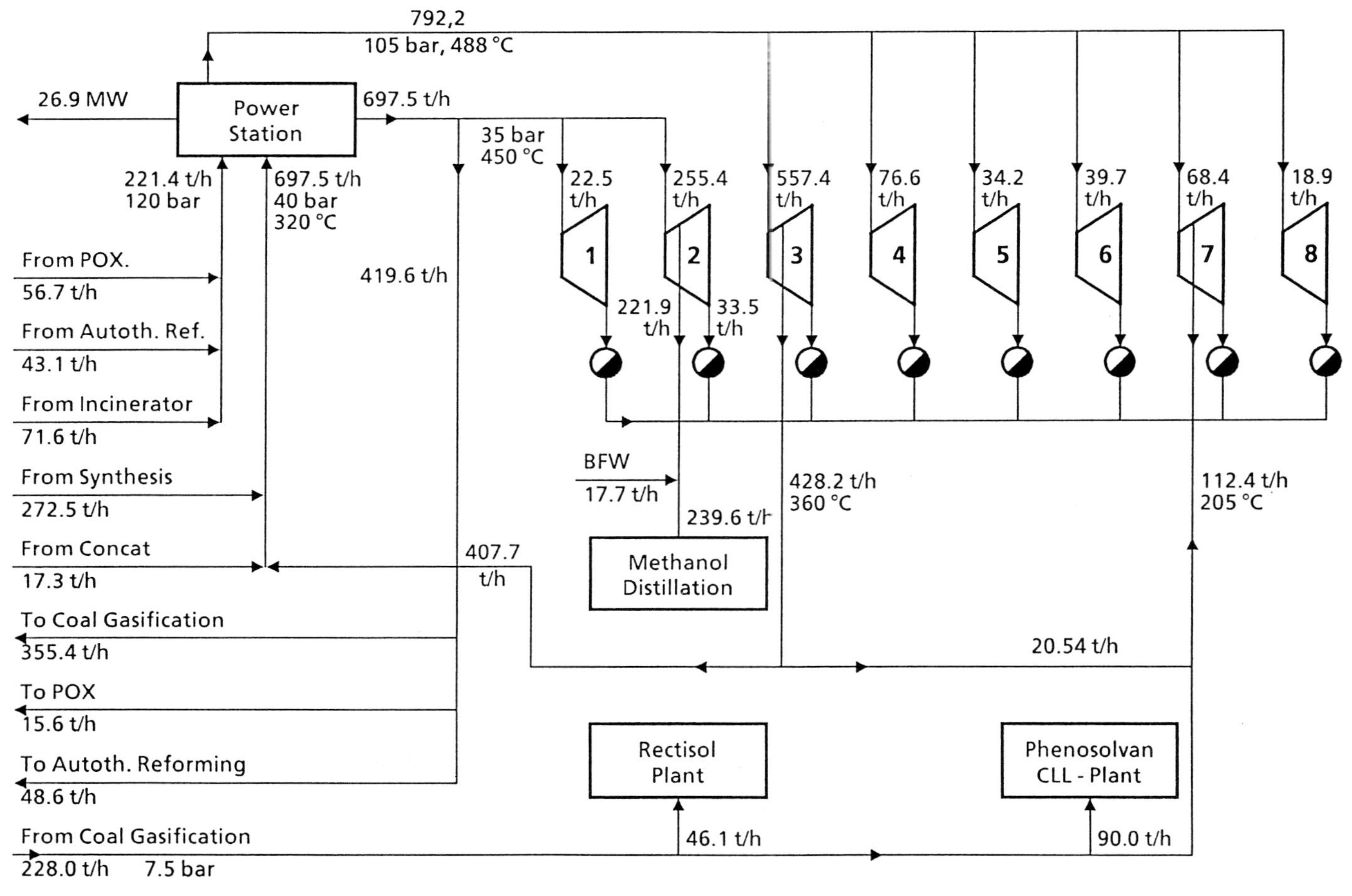

Fig. 7.4. Steam/energy diagram. (1) Recycle compr.; (2) syngas compr.; (3) air compr.; (4) O_2 compr.; (5) refrigeration compr.; (6) PSA off gas compr.; (7) power generator; (8) boiler feed water pump

As described already in Chap. 6, instrument air and compressed air for operating purposes are normally supplied by the air compressor in the air separation unit. A receiver in the instrument air system ensures a continuous supply during malfunctions for a period of 30 minutes. A diesel-powered auxiliary compressor is usually installed to start up the plant and ensure compressed air supply during downtimes.

8. Future Outlook

Today – at the threshold to the nineties – there are only 4 plants in the world which produce methanol from coal. Two of them, one in South Africa and the other in the USA, have a low capacity. The other two – one again in the USA, the other in Germany – have outputs of 650 and 1 200 tons of methanol per day. One of the reasons why only less than 4 % of world methanol production are derived from coal is the fact that coal-based methanol plants of today's size require a great deal more capital than natural-gas-based ones. They are also more complex – as shown in this book – and require more personnel and maintenance.

Actually, the argument that the capital investment in coal-based plants is higher holds true without qualification only where methanol can be produced from either local coal or local natural gas. A comparison between the cost of a coal-based methanol plant in, say, Tennessee/USA and a gas-based plant in South East Asia, e.g. on Bunyu Island/Indonesia, will quickly reveal that the expenditure on infrastructure in the less developed parts of the world may increase the plant cost by 50–100 %, and if the high transportation costs of 25–30 US$ per ton are added to the methanol production price, coal as a raw material can be overlooked in certain parts of the world only as long as the natural gas price remains at its current extremely low level.

A number of reasons suggest that coal will play a much greater role in methanol production during the years to come.

- All evidence points clearly in the direction that in the USA the legislative bodies, at least in a number of states, will enforce the use of neat methanol as a motor fuel. For California alone, this would mean that methanol consumption in motor fuel service there would be two and a half times as high as the current total methanol production in the world, and it would only be natural to build methanol plants in the USA whose capacities would be many times larger than today. This would considerably reduce the capital cost per ton of methanol, and the price differential between natural gas and coal would have a much greater effect on methanol production costs than with today's plant sizes.
- Since coal contains more carbon than required for methanol production, the selection of a suitable gasification process might ensure that methanol could be produced in an ideal combination with high-CO gases – a route which would be attractive above all for the production of acetic acid from methanol and CO, as shown by the above-mentioned plant in the USA.

- As the production route from coal via methanol to automotive fuels has an edge on the coal-based Fischer-Tropsch process not only because of its high selectivity but also in view of its higher efficiency, there are today signs in South Africa – which possesses some of the richest coal deposits in the world – that methanol will play a role as an intermediate or finished product in future automotive fuel production.

The chapters of this book should clearly show that the industrial world is well equipped technologically to act if the above expectations begin to materialize.

References

Chapter 1

1.1 Rohstoff Kohle. Chemie, Weinheim, New York 1978
1.2 Survey of Energy Resources. 11. World Energy Conference, Sep., 1980
1.3 G. B. Fettweiß: Weltkohlevorräte. Glückauf, Essen, 1976
1.4 H. Juentgen, K. H. van Heek: Kohlevergasung. Karl Thiemig, München 1981
1.5 H. G. Franck, A. Knop: Kohleveredlung. Springer, Berlin 1979
1.6 Ullmanns Enzyklopädie der techn. Chemie. Urban und Schwarzenberg, München, Bd. 10, 1958
1.7 The British Gas/LURGI Gasifier, Lurgi Brochure 11/86
1.8 Ullmanns Enzyklopädie der techn. Chemie. Chemie, Weinheim, Bd. 3, 1973
1.9 Coal Gasification Systems. Synthetic Fuels Associates, Mountain View, California, AP-3109 Research
1.10 J. Ruer, F. Bögner: Kohlevergasung im Fluidatbett. B.W.K. 28, 1976, 2
1.11 F. Sabel: Die Winkler'sche Wirbelschichttechnik. Chemie-Ing.-Technik, 24. Jahrg. 1952/2
1.12 H. Ilgner, G. Fabian, F. Honolke: International Coal & Gas Conversion Conference, Pretoria/RSA, 24.8.1987
1.13 J. Lambertz, N. Bruengel, W. Ruddeck, L. Schrader: 2. EPRI Conference Synthetic Fuels – Status and Directions. San Francisco/USA, April 1985
1.14 F. J. Trogus, A. B. Krewinghaus, R. T. Perry: Status of the Shell Gasification Demonstration Plant. 8. EPRI Coal Gasification Conference, Palo Alto/USA, October 1988
1.15 W. L. Heitz et al.: Applicability of the Shell Gasification Process (SCGP) to Low Rank Coals. AICHE Winter National Meeting, Atlanta/USA, March 1984
1.16 T. Matsunara: Texaco Coal Gasification Technology. 69. Regular Conference on Flotation, Ube City, Japan, Nov. 1982
1.17 R. H. Fisackerly, D. G. Sundstrom: The Dow Syngas Project. Publication by Dow Chemical Company, 1986
1.18 J. P. Henley, D. G. Sundstrom, "Initial Experience of the Commercial Dow Coal Gasification Plant", Publication by Dow Chemical Company, 1987
J. Meunier: Vergasung fester Brennstoffe und oxidative Umwandlung von Kohlenwasserstoffen. Chemie, Weinheim, 1962

Chapter 2

2.1 Ullmanns Enzyklopädie der techn. Chemie. Chemie, Weinheim, 2, 575–599, 1972
2.2 Coal Handbook. E.R.A. Meyers, New York, Marcel Dekker 642–645,1981
2.3 W. Herbert: Erdöl u. Kohle 9 (2) 77–81, 1956
2.4 Ullmanns Enzyklopädie der techn. Chemie. Chemie, Weinheim, 4, 432–435, 1977
2.5 M-PYROL Handbook. GAF-Corp., New York
2.6 US Patents, 3,120,993 and 3,324,627
2.7 H. U. Kohrt, K. Thormann, K. Bratzler: Erdöl u. Kohle 2 (16) 96–99, 1963
2.8 US Patent, Serial No. 3,966,875
2.9 R. D. Stoll, S. Roper: Erdöl u. Kohle 3 (35) 380–385, 1982
2.10 D. K. Judd: Hydrocarbon Processing 4, 122–124, 1978
2.11 British Patent Spec. 1 574 646

2.12 US Patent, 2,926,751-3
2.13 S. Franckowiak, E. Nitschke: Hydrocarbon Processing 49, 145–148, 1970
2.14 C. Wehner et al.: Chem. Technik 34 (6) 291–294, 1982
2.15 K. F. Butwell, D. J. Kubek, P. W. Sigmund: Hydrocarbon Processing 3, 108–116, 1982
2.16 Hydrocarbon Processing 4, 81–82, 1971
2.17 Fang-yuan Jou, A. E. Mather, F. D. Otto: Int. Chem. Eng. Process Des. Dev. 21, 539–544, 1982
2.18 Hydrocarbon Processing 4, 100, 1982
2.19 H. N. Weinberg, et al.: 11. World Petroleum Congress, London, 1983
2.20 D. W. Savage et al.: Ind. Chem. Eng. Symposium, Series 104, August 1987
2.21 C. L. Dums et al.: Hydrocarbon Processing 44 (4) 137–140, 1965
2.22 F. C. Riesenfeld, A. L. Kohl: Gas Purification, 2. Edition, 433–451
2.23 S. D. Beskow, Techn Chemische Berechnungen. Technik, Berlin, 532, 1953
2.24 G. Schlinger: Energy Research 4, 127–136, 1980
2.25 L. W. ter Haar: Het Ingenieursblad 42, Nr. 21, 1973
2.26 B. H. Mink: Symposium for Chemical Industry, Berlin, November 1984
2.27 L. J. van Aube: Hydrocarbon Processing, Sept. 1980

Chapter 3

3.1 Emerging Energy and Chemical Applications of Methanol. The World Bank, April 1982
3.2 Methanol. Chemical Economics Handbook, SRI International, Oct. 1983
3.3 R. W. Kirer et al.: J. Chem. Eng. Data 6 (3) 338–341, 1961
3.4 T. Nitta et al.: Chem. Eng. Sci. 29, 2213–2218, 1974
3.5 I. Nagata: J. Chem. Eng. Data 14 (4) 419–420, 1969
3.6 E. Strömsöe, H.G. Rönne, A. L. Lydersen: J. Chem. Eng. Data 15 (2) 286–290, 1970
3.7 C. Carr, J. A. Liddick: Ind. Eng. Chem. 43, 693, 1951
3.8 Ullmanns Enzyklopädie der techn. Chemie. Chemie, Weinheim, 16, 621–623, 1978
3.9 T. W. Yergovich, G. W. Swift, F. Kurata: J. Chem. Eng. Data 16 (2) 222–226, 1971
3.10 H. K. Ross: Ind. Eng. Chem. 46, 607, 1954
3.11 W. J. Thomas, S. Portalsky: Ind. Eng. Chem., 50 (6), 967–970, 1958
3.12 D. D. Wagmann, J. E. Kilpatrick, K. S. Pitzer u. F. D. Rossini: J. Res. Nath. Bur. Standards 35, 467–496, 1946
3.13 S. D. Beskow: Techn. Chem. Berechnungen. Technik, Berlin, 535, 1953
3.14 D. H. Bolton: Chem.-Ing. Techn. 41 (3) 129–134, 1969
3.15 A. Liebgott, W. Herbert u. G. Baron: Brennstoff-Chem. 25 (2) 75–80, 1972
3.16 O. A. Hougen and K. M. Watson: Ind. and Eng. Chemistry 35 (5) 529–541, 1943
3.17 Natta, G., Pino, P., Mazzanti, G. and Pasquon: J. Chim. e Ind. (Milano) 35, 705, 1953
3.18 P. Villa et al.: Ind. Eng. Chem. Process Des. Dev. 24, 12–19, 1985
3.19 H. H. Kung: Catal. Rev., Sci. Eng. 22 (2) 235–259, 1980
3.20 Alcohol Fuels. Symposium, Sydney, Aug. 1978
3.21 R. E. Smith, G. C. Humphreys and G. W. Griffiths: Symposium of 100. Anniversery of Luigi Cassale, Rapallo, Italy, Nov. 1982
3.22 U.S. Patent 4,369,255
3.23 Chem. Economy and Eng. Rev. 13 (6) 17–25, 1981
3.24 Dr. Ib. Dybkjaer: C.E.E.R. 6 (11) 21–33, 1974
3.25 M. B. Sherwin, M. E. Frank: Hyrocarbon Processing 122ff, 11, 1976
3.26 J. Haggin: C & EN 19 (7) 41–42, 1982
3.27 35. Chemical Eng. Conference, Calgary, Oct. 1985
3.28 U. R. Westerterp et al.: Hydrocarbon Processing, 69–73, 11, 1988
3.29 Australian Patent Application A 4–A 1–74 391/81
3.30 German Patent Application P 36 41 774.2, 12/1986
3.31 P. Courty, J.-P. Arlie, A. Convers, P. Mititenko, A. Sugier: L'actualité chimique 11, 19–22, 1983
3.32 R. Ricci, A. Paggini, V. Fattore, F. Ancilotti, M. Sposini: Seminar on Chemicals from Methanol, United Nations, Genova, June 1983
3.33 E. Supp: AICHE Spring National Meeting, New Orleans, LA, April 1986
3.34 Th. O. Wentworth, D. F. Othmer, CEP, 8, 29–35, 1982 SRI International, Oct. 1983

Chapter 4

4.1 N. J. Macnaughton, A. Pinto, P. L. Rogerson: Development of Methanol Technology for Future Fuel and Chemical Markets, AICHE Spring National Meeting, Anaheim, Cal., May, 1984

4.2 E. Supp: AICHE Spring National Meeting, New Orleans, LA, April, 1986

4.3 P. Courtey, J.-P. Arlie, A. Convers, P. Mititenko, A. Sugier: L'actualité chimique 11, 19–22, 1983

Chapter 5

5.1 Bataafse International Petroleum Maatschappij: Deoiling Industrial Waste Water, Design and Operation of the CPI. MF Report No. 68 700

5.2 Ullmanns Enzyklopädie der techn. Chemie. Chemie, Weinheim, 6, 417–463, 1982

5.3 Bundesimmissionsschutzgesetz, Techn. Anleitung zur Reinhaltung der Luft (TA Luft), FRG 1985

5.4 Clean Air Act of 1970, Amendment of 1977, USA

5.5 Ullmanns Enzyklopädie der techn. Chemie. Chemie, Weinheim 21, 2–31, 1982

5.6 B. W. Gamson, R. H. Elkins: Chem. Eng. Progr. 49, 302–315, 1953

5.7 R. S. Coward, J. G. Barron: Gas Conditioning Conference, 1983

5.8 B. G. Goar: 57. Annual GPA Convention, March, 1978

Conversion from Metric to US Units

Metric System	US Units
m	3.281 ft
m^2	10.764 ft^2
m^3	35.315 ft^3
m^3 (liqu.)	264.173 gal
m^3 (gas)	37.230 scft (60°F,30 mm Hg)
m^3/h (Liqu.)	4.403 gal/min
kmol	2.2046 lbmol
kg	2.2046 lb
t	1.1023 st
J	$0.9478 \cdot 10^{-3}$ BTU
kWh	3412.7 BTU
kcal/m^3	0.1066 BTU/scft
kJ/m^3	0.0255 BTU/scft
kJ/kg	0.4299 BTU/lb
bar	14.503 psi
Torr	0.0193 psi
x°C	(x-32)/1.8 °F
xK(Kelvin)	(x-32)/1.8+273 °F

Decimal Prefixes

μ	10^{-6}
m	10^{-3}
c	10^{-2}
d	10^{-1}
k	10^{3}
M	10^{6}
G	10^{9}

M. B. Hocking

Modern Chemical Technology and Emission Control

1985. XVI, 460 pp. 152 figs.
ISBN 3-540-13466-2

Contents: Background and Technical Aspects of the Chemical Industry. – Air Quality and Emission Control. – Water Quality Emission Control. – Natural and Derived Sodium and Potassium Salts. – Industrial Bases by Chemical Routes. – Electrolytic Sodium Hydrocide and Chlorine and Related Commodities. – Sulfur and Sulfuric Acid. – Phosphorus and Phosphoric Acid. – Ammonia, Nitric Acid and their Derivatives. – Aluminium and Compounds. – Ore Enrichment and Smelting of Copper. – Production of Iron Steel. – Production of Pulp and Paper. – Fermentation Processes. – Petroleum Production and Transport. – Petroleum Refining. – Formulae and Conversion Factors. – Subject Index.

This book of applied chemistry considers the interface between chemistry and chemical engineering, illustrated by examples from some of the important process industries. Integrated in this is a detailed consideration of measures which may be taken for a voidance or control of potential emissions.

Springer-Verlag Berlin
Heidelberg New York London
Paris Tokyo Hong Kong